Research Progress in Oligosaccharins

Heng Yin • Yuguang Du
Editors

Research Progress in Oligosaccharins

 Springer

Editors

Heng Yin
Chinese Academy of Sciences
Dalin Institute of Chemical Physics
Dalian City, China

Yuguang Du
Chinese Academy of Sciences
Dalin Institute of Chemical Physics
Dalian City, China

ISBN 978-1-4939-8064-2   ISBN 978-1-4939-3518-5   (eBook)
DOI 10.1007/978-1-4939-3518-5

This Springer imprint is published by Springer Nature
The registered company is Springer Science+Business Media LLC New York

Contents

Contents

Contributors

Peter Albersheim, Ph.D. Department of Biochemistry and Molecular Biology, Complex Carbohydrate Research Center, University of Georgia, Athens, GA, USA

Guochuang Chen, Ph.D. School of Life Science and National Glycoengineering Research Center, Shandong University, Jinan, China

Kaoshan Chen, Ph.D. School of Life Science and National Glycoengineering Research Center, Shandong University, Jinan, China

Department of Pharmacy, Wannan Medical College, Wuhu, China

State Key Laboratory of Microbial Technology, Shandong University, Jinan, China

Yoshitake Desaki, Ph.D. Department of Life Sciences, School of Agriculture, Meiji University, Kawasaki, Kanagawa, Japan

Yuguang Du, Ph.D. Biotechnology Department, Dalian Institute of Chemical Physics, Chinese Academy of Sciences, Dalian, Liaoning, China

Institute of Process Engineering, Chinese Academy of Sciences, Beijing, China

Moran Guo, Ph.D. Shijiazhuang Center for Disease Control and Prevention, Shijiazhuang, China

Marcello Iriti, Ph.D. Department of Agricultural and Environmental Sciences, Milan State University, Milan, Italy

Shuguang Li, B.S. Biotechnology Department, Dalian Institute of Chemical Physics, Chinese Academy of Sciences, Dalian, China

Alejandra Moenne, Ph.D. Department of Biology, Faculty of Chemistry and Biology, University of Santiago of Chile, Santiago, Chile

Naoto Shibuya, Ph.D. Department of Life Sciences, School of Agriculture, Meiji University, Kawasaki, Kanagawa, Japan

Tomonori Shinya, Ph.D. Department of Life Sciences, School of Agriculture, Meiji University, Kawasaki, Kanagawa, Japan

Institute of Plant Science and Resources, Okayama University, Okayama, Japan

Elena Maria Varoni, Ph.D. Dental Unit II, Department of Biomedical, Surgical and Dental Sciences, San Paolo Hospital, Milan State University, Milan, Italy

Wenxia Wang, Ph.D. Biotechnology Department, Dalian Institute of Chemical Physics, Chinese Academy of Sciences, Dalian, China

Heng Yin, Ph.D. Biotechnology Department, Dalian Institute of Chemical Physics, Chinese Academy of Sciences, Dalian, China

Xiaoming Zhao, Ph.D. Biotechnology Department, Dalian Institute of Chemical Physics, Chinese Academy of Sciences, Dalian, China

Chapter 1
The Discovery of Oligosaccharins

Peter Albersheim

Abstract This chapter is a personal historical account of the events leading to the discovery of oligosaccharins. The discovery was not the result of a eureka-type event but rather lots of well-designed laboratory experiments. In the early days, including most of the time period covered by this chapter, oligosaccharins were called elicitors. The chronology of the chapter ends in 1984, for by that time the biological regulatory properties of structurally defined oligosaccharides were established. The biological activities and structures of two oligosaccharins originating from fungal mycelial wall polysaccharides and two from plant cell wall polysaccharides are described.

Keywords Oligogalacturonides • Chitin • Chitosan • β-Glucan • Elicitor and oligosaccharin

Introduction

I hesitated when invited to write this chapter because it is difficult to write about myself. I decided to accept the offer and to use it to describe events that influenced me to become a scientist and acknowledge a few of the people who were participants in the discovery of oligosaccharins. Teams move bioscience forward most rapidly. The uncovering of oligosaccharins is no exception.

It is now widely known that complex carbohydrates (those with more than one sugar residue) are involved in many if not most biological processes. Thus it is no longer surprising that some oligosaccharides are biologically active; that is, they are regulatory molecules, with the ability to turn on or turn off processes at concentrations as low as nano molar. **Such biologically active oligosaccharides are called oligosaccharins**.

P. Albersheim, Ph.D. (✉)
Department of Biochemistry and Molecular Biology, Complex Carbohydrate Research Center, University of Georgia, 1575 Morton Road, Athens, GA 30605, USA
e-mail: palbersheim@ccrc.uga.edu

© Springer Science+Business Media New York 2016
H. Yin, Y. Du (eds.), *Research Progress in Oligosaccharins*,
DOI 10.1007/978-1-4939-3518-5_1

Oligosaccharins were discovered in the 1970s and 1980s, when most biochemists did not give much thought to carbohydrates. The vast majority of research biochemists targeted nucleic acids and proteins. Textbooks for general biochemistry courses gave little attention to complex carbohydrates focusing in plants on cellulose, starch, and sucrose. Carbohydrates were also known to be components of glycoproteins, glycolipids, and nucleic acids, but this knowledge did not cause any member of our research team, or apparently anyone else, to suggest that oligosaccharides could function much like hormones.

Early Influences

My odyssey started at the age of 7 when my family moved from Long Island, N.Y., to the New Jersey shore. On the bus taking me to the first day in my new school I met Jim Carton, another 7-year-old who lived near me. Together we came to love the outdoors. Many after-school hours were spent fly-fishing in the lake that surrounded our community. In the winter we trapped muskrats, often riding our bikes for miles in the cold blackness of early mornings in order to service our traps before school. My large German shepherd provided protection from bullies who threatened our traps and bounty.

My first experience living on a farm came in the summer I was 10. I liked it. When I was 15 my family took a long vacation driving from New Jersey to California and back with appropriate stops. I had plenty of opportunity to see farms and ranches, which together with the fly-fishing opportunities and beautiful mountains impressed me to the point where I decided I would learn about farming or ranching with the goal of living in the west.

Plant Pathology and Chemistry

I worked on farms for three summers starting the summer after I graduated from high school and after each of my first 2 years at the Cornell University School of Agriculture. The farming experience has impacted my life in numerous ways. Perhaps the most important for this chapter was my awakening to the significance of plant diseases. I became aware of their economic and social impact on farm families, which could be devastating.

I noticed in a corn field, essentially destroyed by a fungal infection, several plants standing taller than those around them. I examined those plants and saw that they were healthy corn plants, apparently oblivious to the fungus that surrounded them. I wondered what endowed these plants with resistance to the fungus and found the answer wasn't known. It impelled me to major in plant pathology. Some 10 years later I began to do research on host-pathogen interactions, research that led to oligosaccharins.

Agricultural economics was the subject of one of my first-year courses at Cornell. I learned of the tremendous cost of acquiring a farm and realized that I would most likely have to work for someone else if I stayed in agriculture. Since being independent was one of the attractions of farming, I became disillusioned with the idea. Fortunately I had an alternative. I took chemistry courses in the Arts and Sciences College at Cornell equivalent to that of a major in chemistry. That gave me the freedom to decide what to do next and I chose to continue my education in science. I enrolled in graduate school in the California Institute of Technology in September of 1956. This came after my friend Jim Carton and I had a fabulous 3-month cross-country vacation with lots of fly-fishing, camping, sightseeing, and girl meeting.

Research During Graduate School and Postdoctoral

My first day at Caltech I was assigned a desk in a two-person research lab. My lab mate was a graduate student who had come from Cornell 2 years earlier. My second day at Caltech I assisted my lab mate with an experiment—the first research experience of my life. I fell in love with research and thereafter never doubted that I had made the right choice by committing myself to a career in science.

My thesis advisor was James Bonner, a well-known plant physiologist who was particularly interested in how the hormone auxin (indoleacetic acid) stimulates plant cells to elongate. Plant cells are encased in a wall that defines the shape and size of the cell. The thickness of the wall does not change appreciatively as the cell elongates. Thus Bonner reasoned that new material must be added to the wall and auxin probably stimulates the incorporation into the wall of the new material. My thesis provided evidence in support of Bonner's hypothesis. It was known, at the time, that the walls of growing cells are composed of cellulose, pectin, and additional uncharacterized polysaccharides. Pectin's structure was not known but it was known to be rich in galacturonic acid, an acidic sugar (Fig. 1.1). I showed that auxin does in fact stimulate the incorporation of pectin into cell walls. At the very beginning of my research career I had come face to face with carbohydrate chemistry.

These experiences with complex carbohydrates combined with my interest in plant pathology were two of the factors that led eventually to oligosaccharins. A third factor was my experience with enzymes that cleave polysaccharides. I had a stroke of luck in the spring of 1959 during a 3-month postdoctoral at Caltech. I was looking for a new pectin-cleaving enzyme using as substrate a viscous solution of citrus pectin in pH 6.8 buffer. I boiled the solutions to stop the reactions and to my surprise the solution in every reaction lost its viscosity including control solutions with no enzyme. Boiling for 5 min at pH 6.8 unexpectedly depolymerized the pectin. I found out, during a following postdoctoral year at the Swiss Federal Institute of Technology, that I had stumbled on the lyase reaction (Fig. 1.1). During that postdoctoral I found pectin lyase, a new type of pectin-cleaving enzyme that functions at physiological temperatures the same way boiling at pH 6.8 does.

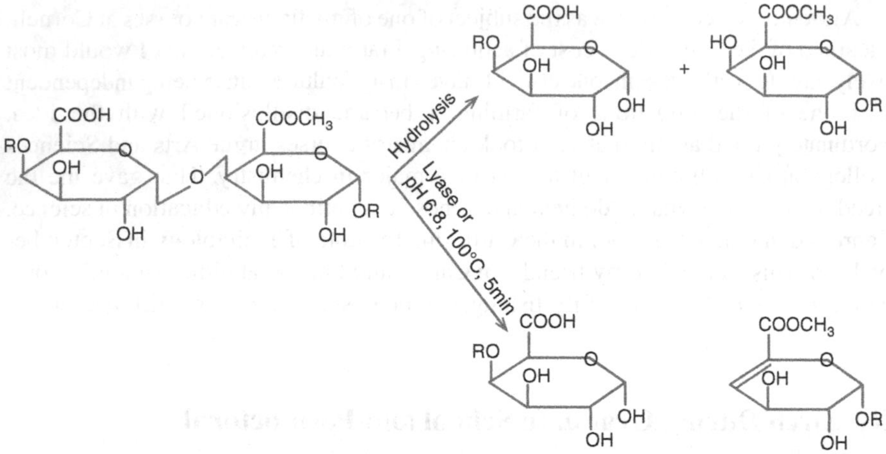

COOH COOCH₃

RO / OH / O / OH / OR / OH / OH

Hydrolysis

Lyase or pH 6.8, 100°C, 5min

COOH COOCH₃

RO / OH / OH / OH + HO / OH / OR / OH

COOH COOCH₃

RO / OH / OH / OH / OH / OR / OH

Fig. 1.1 Polygalacturonic acid is made up of a chain of galacturonic acid residues, some of which are methyl esterified. R designates the chain. There are two ways to break a bond between two galacturonic acid residues. One is by hydrolysis that adds water molecule to the bond and the other way is by a lyase reaction. Both reactions can be catalyzed by enzymes. The lyase reaction has a second way to be catalyzed and that is by boiling at pH 6.8 for 5 min. The boiling only works on a galacturonic acid residue that is methyl esterified. A lyase enzyme can work whether or not a galacturonic acid residue is methyl esterified

My Own Lab—Cell Walls

These findings won me an appointment as a Lecturer at Harvard, which after 1 year was converted to an Assistant Professorship. A graduate student, Gretchen Becker, a postdoctoral, Paul Hui, and myself made the most pertinent scientific contribution of my small group during my 4-year stint at Harvard. We imported suspension-cultured sycamore cells from Derek Lamport of Cambridge University. We were able to grow the cells and show that the cells secreted polysaccharides. Later we showed that suspension-cultured sycamore cells as well as suspension-cultured cells of many plants secrete intact cell wall polysaccharides into the culture medium. This enabled us to purify and structurally characterize the polysaccharides without fragmenting them in order to extract them from walls. One reason it is important to know the structures of the wall polysaccharides is because they are a source of oligosaccharins.

I have been fortunate to have a great team of coworkers, many of whom contributed to the structural analysis of cell wall polysaccharides. Three graduate students, Ken Keegstra, Dietz Bauer, and Ken Talmadge, worked as a team; each took a polysaccharide to work on. They contributed enormously to that project. Before they finished their PhDs we had a structural model of the primary cell wall.

Bengt Lindberg at the University of Stockholm, Sweden, headed one of the important groups in the 1960s structurally characterizing complex carbohy-

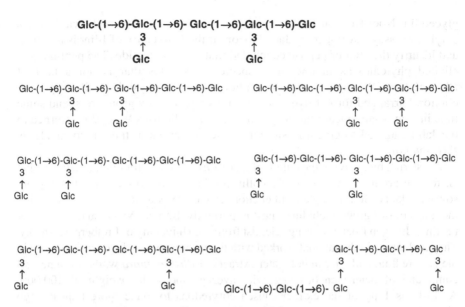

Fig. 1.2 This figure depicts seven heptaglucoside fragments generated by partial acid hydrolysis of beta glucan. Beta glucan is a polysaccharide present in mycelial walls of fungi. Only one heptaglucoside out of 300 generated by acid hydrolysis is an active oligosaccharin. The active fragment is presented here on the top of the figure in *bold*. All the residues are D-glucosyl (Glc) and all of them are in beta form. This figure shows the exquisite structural specificity required of the active fragment

drates. I visited his laboratory to learn their methods. They were able to determine the primary structures of polysaccharides with repeating glycosyl residue sequences of up to seven residues. We imported their technology and added a new procedure that enabled us to obtain the structures of polysaccharides with repeats as large as 11 glycosyl residues. Mastering this technology is a major reason we were able to decipher the structure of the heptaglucoside oligosaccharin (Fig. 1.2).

The Heptaglucoside Oligosaccharin

Our team entered the host-pathogen interaction field by extending elicitor studies of other scientists. Elicitors include molecules that, when applied to plants, cause plants to act defensively. Perhaps the most well-studied plant defense reaction is their ability to synthesize phytoalexins. Phytoalexins are antibiotics made by plants at the site of attempted infections. One system that had been studied is the ability of autoclaved mycelia of the fungal pathogen of soybean, *Phytophthora megasperma glycenia*, to cause soybean tissues to make the phytoalexin glyceollin. Noel Keen at the University of California at Riverside was using a cotyledon assay developed by Jack Paxton of the University of Illinois to purify and identify the elicitor [1]. He concluded that it was a peptide. The peptide only elicited phytoalexins in one plant species, a species that is not a host of *Phytophthora*. We thought that the peptide was not a physiologically important elicitor. I arranged for Art Ayers, a graduate student in my group, to spend some time in Noel's lab to learn the cotyledon assay for elicitors. When Art returned to our lab he agreed to give a fresh look at the properties of the elicitor made by *Phytophthora*.

Art started by assaying the culture fluid of *Phytophthora* for elicitor activity, that is, for the ability to stimulate glyceollin production in cotyledons of 8-day-old soybean plants. He found plentiful elicitor activity. Art was joined by several other members of our group including, most importantly, Barbara Valent and, for part of the time, Juergen Ebel, a visiting scientist from the University of Freiberg, Germany. They and others in the lab that worked with them showed that the elicitor activity in the culture fluid and later in hot water extracts of *Phytophthora* walls was a neutral compound of heterogeneous size with average molecular weight of 100,000. As little as 1 µg of the extract caused cotyledons to make more than enough glyceollin to stop the growth of *Phytophthora*. We were able to get even more elicitor activity out of mycelial wall extracts by partial acid hydrolysis of the hot water-extracted material. Using a high-resolution sizing column combined with mass spectrometric analysis, we were able to show that the smallest active fragment was seven glucosyl residues and was, in fact, a fragment of beta-glucan, the major polysaccharide component of *Phytophthora* mycelial walls. We found by methylation analysis that the fragments were rich in terminal, 3-, 6-, and 3,6-linked glucosyl residues. We correctly proposed that the smallest elicitor was a 6-linked pentaglucoside with terminal glucosyl residues attached to C-3 of two of the 6-linked residues [2]. We thought that we would confirm the structure soon. We were wrong. It took another 8 years before we could publish the primary structure of the elicitor.

The problem was that partial acid hydrolysis of the mycelial-wall beta-glucan forms about 300 heptaglucosides and only one of these has elicitor activity. Since all these heptaglucosides are neutral molecules and all of them have the same number of hydroxyl groups, it was very difficult to separate them. A graduate student, Jan Sharp, persevered and finally purified the active fragment by reverse-phase column chromatography. We did not predict the high-resolving power of hydrophobic bonds in molecules like the heptaglucosides that are loaded with hydrophilic hydroxyls, but the heptaglucosides have different affinities to a reverse-phase column and are differentially eluted by 2 % acetonitrile in water.

Jan not only purified the elicitor-active heptaglucoside but she also purified six inactive heptaglucosides that were structurally closely related to the active fragment. Jan then determined the primary structures of all seven heptaglucosides (Fig. 1.2) [3]. Furthermore, a team at the University of Stockholm, under the direction of Per Garegg, chemically synthesized the oligosaccharin-active hepta-glucoside, and we showed that the synthetic heptaglucoside has the same fantastically high activity as the natural product. About nano gram is all that is needed to elicit soybean tissues to synthesize phytoalexins. These results confirmed the structure of the oligosaccharin and the exquisite structural specificity required for activity [4].

When we started looking for an elicitor of phytoalexins we thought that we would need a gram of pure elicitor to succeed and, after obtaining pure elicitor, it might take us up to 10 years to determine its structure. Because of technological advances for determining the structures of small amounts of complex carbohydrates, when we finally purified the oligosaccharin it required about 1 week to determine its structure and we only had about 50 μg of pure elicitor to work with [3].

The Second Oligosaccharin from a Mycelial Wall Polysaccharide

Lee Hadwiger and his coworkers at Washington State University in Pullman, Washington, found that chitosan elicits phytoalexin production in pea pods [5]. Chitosan is a polysaccharide derived from chitin. Chitin, a component of mycelial walls, is a linear beta-1,4-linked polymer of N-acetyl glucosamine. Hadwiger's group used partial acid hydrolysis to remove the N-acetyl groups and partially depolymerize chitin, thereby forming chitosan. The chitosan was active as an oligosaccharin at about micro molar concentrations, which is pretty good considering that it was an impure preparation.

The First Oligosaccharin from a Plant Cell Wall Polysaccharide

In the meantime work was going forward in two laboratories on another oligosaccharin. Charles West and his group at the University of California in Los Angeles found that an enzyme that cleaves polygalacturonic acid elicits phytoalexins in castor bean [6]. While Michael Hahn, a graduate student in our group, was showing that an oligogalacturonic acid fragment of a plant cell wall polysaccharide elicits phytoalexins in soybean cotyledons [7]. Charles West showed that his enzyme released oligosaccharin-active pectin fragments from the cell wall [8]. We showed that the oligosaccharin is a linear α-1,4-oligogalacturonide that must be between 10 and 14 galacturonic acid residues in length to be active [9]. The oligogalacturonide was the first fragment of a plant cell wall polysaccharide that was identified as an oligosaccharin.

Clarence Ryan and his coworkers at Washington State University in Pullman, Washington, have studied plant proteins that inhibit proteinases secreted by insects and bacteria when they attack plants [10]. The amount of the plant's proteinase inhibitors increases perceptively when the plant is attacked by insects or infected by bacteria. Ryan's group found out that the elicitor of the proteinase inhibitors in tomato seedlings was an oligosaccharin and, in fact, was an oligogalacturonide. However, the size of the active oligogalacturonides required for activity in the tomato system is less stringent than is required for the soybean system.

Oligogalacturonides with 11–14 residues have many other documented effects on plants, perhaps as many as auxin or cytokinin. One effect that fascinates me is the ability of this oligosaccharin to induce tobacco explants to form inflorescences when grown on a medium that would produce no organs in the absence of added oligosaccharin. Furthermore, it requires only micro molar amounts of active oligogalacturonide to induce about five inflorescences to form per explant, a concentration tenfold less than is required for auxin and cytokinin in a flower-inducing medium in the absence of added oligosaccharin [11].

The Second Oligosaccharin from a Plant Cell Wall Polysaccharide

Another oligosaccharide fragment of a plant cell wall polysaccharide was found to be an oligosaccharin. Will York, a graduate student in our group, was extracting hemicelluloses from plant cell walls with alkali when he found that there was some molecule or molecules in the extract that inhibited auxin-induced growth of pea stems. Will knew that the base extract was rich in xyloglucan fragments, the dominant hemicellulose of dicot cell walls. Will treated purified xyloglucan with endoglucanase from a fungal pathogen generating well-defined xyloglucan fragments (Fig. 1.3). Will found that a nine glycosyl-residue fragment inhibits auxin-induced growth of pea stems at concentrations 1000-fold less than the concentration of auxin needed to stimulate the growth of pea stems. A related seven-residue fragment does not inhibit the auxin-stimulated growth [12].

Teamwork Led to the Establishment of the Complex Carbohydrate Research Center

I was fortunate to have an outstanding team of coworkers. I will mention three of them. Alan Darvill came to the group as a postdoctoral for a 1-year stay that turned into a lifetime. Alan became my lieutenant by his leadership qualities and his excellent science. He came in 1975 and within 2 years he became a go-to person for graduate students and postdocs who were having trouble with their research. He not only helped them with their science but also stimulated them to work hard and to

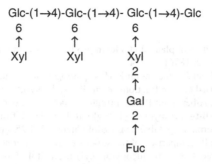

Glc-(1→4)-Glc-(1→4)- Glc-(1→4)-Glc

Fig. 1.3 The nonasaccharide is a fragment of the cell wall polysaccharide known as xyloglucan. It is an active oligosaccharin, while heptasaccharide, which lacks the fucosyl and galactosyl residues of the nonasaccharide, is inactive. The four glucosyl (Glc) residues are all the D enantiomers and beta linked. The three xylosyl (Xyl) residues are all D enantiomers and alpha linked. The galactosyl (Gal) residue is D and beta linked, while the fucosyl (Fuc) residue is L and alpha linked

enjoy life. Barbara Valent joined the lab as a second-year undergraduate student. She stayed for graduate school and a postdoctoral. Barbara contributed tremendously to the research productivity of the lab. Finally I would like to mention Mike McNeil. Mike came as a technician with a Masters degree from MIT. Carbohydrate chemistry was new to him but he became a professional carbohydrate chemist and his work was outstanding. It won Mike a PhD degree from MIT as he had finished all the course work and exams and only had his research to do when he joined our group. Al, Barbara, and Mike joined me in supervising the group, which had grown to some 30 people by 1980.

Carbohydrate science is so demanding, so complex that, even with our large group, we felt the need for colleagues who are specialists in such areas as nuclear magnetic resonance spectroscopy, mass spectrometry, genomics, proteomics, and the like. We needed a center of excellence in carbohydrate science and set out to establish one. We wanted to do it at the University of Colorado but, although the administration approved of the idea, they just did not have the necessary resources. In 1985 Alan and I and 14 other members of the group moved to the University of Georgia in Athens, Georgia, to organize the Complex Carbohydrate Research Center. Alan was the Associate Director of the Center for 2 years when he was promoted to a shared Directorship with me. Since I retired in 2010, Alan has been running the Center by himself and doing a fantastic job. The Center has grown to 17 outstanding tenure-track faculties and several equally outstanding non-tenure track faculties who, together with technicians, graduate students, postdocs, visiting scientists, and office staff, now number close to 300 people. The technology developed there is particularly valuable for oligosaccharin research.

References

1. Keen NT. Specific elicitors of plant phytoalexin production: detenninants of race specificity in pathogens? Science. 1975;187(4171):74–5.
2. Ayers AR, Valent B, Ebel J, Albersheim P. Host pathogen interactions: XI. Composition and structure of wall-released elicitor fractions. Plant Physiol. 1976;57:766–74.
3. Sharp JK, McNeil M, Albersheim P. The primary structures of one elicitor-active and seven elicitor-inactive hexa(β-D-glucopyranosyl)-D-glucitols isolated from the mycelial walls of Phytophthora megasperma f. sp. Glycinea. J Biol Chem. 1984;259(18):11321–36.
4. Sharp JK, Albersheim P. Comparison of the structures and elicitor activities of a synthetic and a mycelial-wall-derived hexa(β-D-glucopyranosyl)-D-glucitol. J Biol Chem. 1984;259(18): 11341–5.
5. Hadwiger LA, Beckman JM. Chitosan as a component of pea-Fusarium solani interactions. Plant Physiol. 1980;66(2):205–11.
6. Stekoll M, West CA. Purification and properties of an elicitor of castor bean phytoalexin from culture filtrates of the fungus Rhizopus stolonifer. Plant Physiol. 1978;61(1):38–45.
7. Hahn MG, Darvill AG, Albersheim P. Host-pathogen interactions: XIX. The endogenous elicitor, a fragment of a plant cell wall polysaccharide that elicits phytoalexin accumulation in soybeans. Plant Physiol. 1981;66:1161–9.
8. Lee SC, West CA. Properties of Rhizopus stolonifer polygalacturonase, an elicitor of casbene synthase activity in castor bean (Ricinus communis L.) seedlings. Plant Physiol. 1981;67(4):640–5.
9. Spiro MD, Kates KA, Koller AL, O'Neill MA, Albersheim P, Darvill AG. Purification and characterization of biologically active 1,4-linked α-D-oligogalacturonides after partial digestion of polygalacturonic acid with endopolygalacturonase. Carbohydr Res. 1993;247:9–20.
10. Bishop PD, Pearce G, Bryant JE, Ryan CA. Isolation and characterization of the proteinase inhibitor-inducing factor from tomato leaves. Identity and activity of poly- and oligogalacturonide fragments. J Biol Chem. 1984;259(21):13172–7.
11. Marfa V, Gollin DJ, Eberhard S, Mohnen D, Darvill AG, Albersheim P. Oligogalacturonides are able to induce flowers to form on tobacco explants. Plant J. 1991;1(2):217–25.
12. York WS, Darvill AG, Albersheim P. Inhibition of 2,4-dichlorophenoxyacetic acid-stimulated elongation of pea stem segments by a xyloglucan oligosaccharide. Plant Physiol. 1984;75:295–7.

Other References

Albersheim P, Darvill AG, McNeil M, Valent BS, Hahn MG, Lyon G, Sharp JK, Desjardins AE, Spellman MW, Ross LM, Robertsen BK, Aman P, Franzen LE. Structure and function of complex carbohydrates active in regulating plant-microbe interactions. Pure Appl Chem. 1981;53:79–88.

Chapter 2
Analysis and Separation of Oligosaccharides

Wenxia Wang, Shuguang Li, Yuguang Du, and Heng Yin

Abstract Oligosaccharides have been widely applied in plant growth promotion and defense induction. Some research suggests that the activity of oligosaccharides is related to their constitutional monomer structure and degree of polymerization. Thus analysis and separation of oligosaccharides is essential to understand their structure–function relationships in plant science. The basic analysis and separation process is composed of two parts: separation and detection. The most common used for oligosaccharide separation is ion-exchange chromatography and hydrophilic interaction liquid chromatography, as well as the detectors used include refractive index detector, evaporative light-scattering detector, pulsed amperometric detector, and mass spectrometers. A suitable chromatography and detector can be chosen to achieve a satisfactory analysis result according to the structure of oligosaccharide to be analyzed. This chapter mainly reviews recent progresses in the separation and analysis of the oligosaccharides as biostimulants in agriculture, including chitosan oligosaccharide, chitin oligosaccharide, alginate oligosaccharide, oligogalacturonic acid, and carrageenan oligosaccharide. A variety of chromatographic examples are given in this chapter.

Keywords Oligosaccharides • Analysis • Separation • Ion-exchange chromatography • Hydrophilic interaction liquid chromatography

W. Wang, Ph.D. (✉) • S. Li, B.S. • H. Yin (✉)
Biotechnology Department, Dalian Institute of Chemical Physics,
Chinese Academy of Sciences, 457# Zhongshan Road, Dalian, Liaoning 116023, China
e-mail: wangwx@dicp.ac.cn; lsg@dicp.ac.cn; yinheng@dicp.ac.cn

Y. Du
Biotechnology Department, Dalian Institute of Chemical Physics,
Chinese Academy of Sciences, 457# Zhongshan Road, Dalian, Liaoning 116023, China

Institute of Process Engineering, Chinese Academy of Sciences,
1 North 2nd Street, Zhongguancun, Haidian District, Beijing 100190, China
e-mail: ygdu@ipe.ac.cn

© Springer Science+Business Media New York 2016   11
H. Yin, Y. Du (eds.), *Research Progress in Oligosaccharins*,
DOI 10.1007/978-1-4939-3518-5_2

Abbreviations

HILIC	Hydrophilic interaction liquid chromatography
COS	Chitosan oligosaccharide
TLC	Thin-layer chromatography
DP	Degree of polymerization
RI	Refractive index
Mn	Number average molecular weight
Mw	Weight average molecular weight
GPC	Gel permeation chromatography
ELSD	Evaporative light scattering detector
PAD	Pulsed amperometric detector
CAD	Charged aerosol detector
MS	Mass spectrometers
AOS	Alginate oligosaccharides
GlcNAc	N-acetylglucosamine
HPAEC	High-performance anion-exchange chromatography
ESI–MS	Electrospray ionization mass spectrometry
M	Mannuronic acid
G	Guluronic acid
ESI-MS/MS	Electrospray ionization tandem mass spectrometry
CID	Collision-induced dissociation
OGA	Oligogalacturonic acid
SEC	Size-exclusion chromatography
PGC	Porous graphitic carbon
SORI-CID	Sustained off-resonance irradiation-collision-induced dissociation
PSD	Post-source decay

Introduction

Oligosaccharides found in various natural origins have multiple functions in eliciting plant defense responses and promoting plant growth, which received much attention for application in agriculture. Besides their biological research, the analysis and separation of oligosaccharides is also a very important step to understand the mechanism of action which is often related to their constitutional monomer structure and degree of polymerization.

However, the analysis and separation of mixtures of oligosaccharides is not straightforward because of the complex structure of oligosaccharide itself and the lack of available commercial standards. Usually analysis of oligosaccharides is focused on two aspects: separation and detection. At present, the most efficient separation method of oligosaccharides is adsorption or partition chromatography on bonded-phase silica and ion-exchange resins. The most used detectors for oligosaccharide analysis include refractive index (RI) detector, evaporative light-scattering detector (ELSD), UV detector, pulsed amperometric detector (PAD), charged aerosol detector (CAD), and mass spectrometers (MS). Considering the composition and sequence of oligosaccharide to be analyzed, different separation method and detector can be chosen to achieve a satisfactory analysis result. Ion-exchange chromatography coupled with electrochemical detection has been widely employed for oligosaccharide analysis. This separation method exhibits high resolution and sensitivity; the main drawback is the incompatibility with MS. Hydrophilic interaction liquid chromatography (HILIC) was a recently developed powerful technique for oligosaccharide separation, where analytes are separated on a polar stationary phase and eluted by a binary mobile phase with the main component usually being 5–40 % water in acetonitrile. The advantages of HILIC, such as good retention of polar compounds, high selectivity, and MS compatibility, make it one of the most suitable chromatographic modes for oligosaccharide separation.

This chapter mainly focuses on the recent developments in the analysis and separation of oligosaccharides as biostimulants in agriculture by high-performance liquid chromatography. A variety of chromatographic methods are included.

Chitosan Oligosaccharides

Chitosan oligosaccharide (COS) is the fragment of chitosan which is produced by deacetylation of chitin. Chitosan oligosaccharide is a well-known elicitor in plant and has been widely used to mimic pathogen attack and shown to induce plant defense responses. However, it is uncertain whether their eliciting activity depends on the degree of deacetylation or the sequence of residues within the oligomer. Recently, several studies reported the analysis and separation method of COS.

Thin-layer chromatography (TLC) is the most traditional, convenient, direct, and economic chromatographic method. Cabrera et al. separated the chitooligomers with DPs up to six using silica gel plates (MERCK 60. GF-254); however, higher DPs could not be successful separated by this method [1]. Using the TLC method, Maria analyzed hydrolysis products of chitosan produced by crude chitosanase from *Paenibacillus ehimensis*. The TLC profiles showed that chitosan was hydrolyzed to oligomers of GlcN composed of DP 2–6 [2].

The RI detector is a universal detector. The detection principle involves measuring of the change in refractive index of the column effluent passing through the flow cell. It can also be applied for COS analysis. Kitter et al. reported an analysis method by loading the chitosan oligomeric fractions onto Lichrosorb-NH$_2$ column (4.0 × 250 mm) connected to a Shimadzu LC-3A HPLC system (Shimadzu Corp., Kyoto, Japan). Elution was performed with 70:30 acetonitrile–water (v/v) as the mobile phase at room temperature at a flow rate of 0.8 ml/min and using a RI detector [3]. Jeon et al. analyzed the oligosaccharides by HPLC on TSK gel NH$_2$-60 column (4.6 × 250 mm, TOSOH Corp., Japan) and refractive index detector using 60 % acetonitrile as elution buffer and a flow rate of 0.8 ml/min [4]. Qin et al. also characterized the number average molecular weight (Mn) and weight average molecular weight (Mw) of samples by gel permeation chromatography (GPC) and an RI 150 refractive index detector. TSK G3000-PW column was used in this study. The eluent was 0.2 mol/L CH_3COOH/0.1 mol/L CH_3COONa and flow rate kept at 1.0 ml/min. The sample concentration reached 0.4 % (w/v) [5]. Vishu kumar et al. determined the DP of chitosan oligomers by an aminopropyl column (3.9 mm × 300 mm; Waters Corp., USA) combined with an RI detector. Acetonitrile/water (70:30) mixture was the mobile phase used at a flow rate of 1.0 ml/min. HPLC profiles can show the chitosan oligomers of DP 2–6 as well as monomer [6]. Hsiao et al. used an ICI HPLC system (LC1100 pump, Australia) equipped with RI detector to analyze chitosan oligosaccharides. The column is a HYPERSIL HS APS column (25 cm × 4.6 mm Thermo Instrument Inc., USA); acetonitrile and distilled water (60/40) mixture was used as mobile phase with a flow rate of 0.8 ml/min at 40 °C. Heptamer and higher DP oligomers can be separated by this method [7].

ELSD as an evaporative mass detector is suitable for the detection of nonvolatile sample components in a volatile eluent. It is also a universal detector available to oligosaccharides. Dong et al. used a weak cation-exchange column with ELSD to separate COS. The Xcharge WCX column (150 × 4.6 mm, Acchrom, China) was used in this study. Through optimization of the HPLC condition, acetonitrile and distilled water (30/70) mixture with 100 mM ammonium acetate was chosen as mobile phase. The chitosan oligomers of DP 2–8 can be well separated under such condition [8].

PAD is commonly used to determine the analytes with electrochemical activity. The hydroxyl groups in oligosaccharides can be oxidized on a working electrode surface and the resulting current can be measured. Tokuyasu et al. used HPLC on a Dionex DX-300 chromatograph equipped with a PAD (Dionex Corp., USA) to monitor the deacetylation reaction of chitin oligosaccharides. A prepacked column of Dionex CarboPacTM PA1 (4 × 250 mm) and a CarboPac TM PA1 guard column (4 × 50 mm) were used in their procedure, and the mobile phase was 30 mM NaOH; the column was washed with 100 mM NaOH and a mixture of 1 M NaOAc and 100 mM NaOH after each analysis, and the flow rate was 1.0 ml/min [9].

CAD can detect any nonvolatile or semivolatile analytes with or without a chromophore. As a universal detector, it can give a consistent response apart from their properties. Therefore, using CAD to analyze oligosaccharides seems to be a good choice. An efficient HILIC combined with CAD detector has been developed for separation of COS in our lab. Nonlinear gradient was used for the "click" Xamide column (250 × 4.6 mm) with acetonitrile as mobile phase A, and ammonium formate solution (100 mM, pH 3.0) as mobile phase B. The gradient is as follows: 0–15 min: linear from 36 % B to 41 % B, 15–60 min: linear from 41 % B to 46 % B, 60–65 min: hold 90 % B. The flow rate is 1 ml/min. Under the eluent conditions, the separation of COS was achieved and their DPs ranged from 2 to 16 (Fig. 2.1).

MS is a very popular and versatile method for COS analysis. Cabrera et al. analyzed the composition of the enzymatic hydrolyzed fractions of chitosan using MALDI-

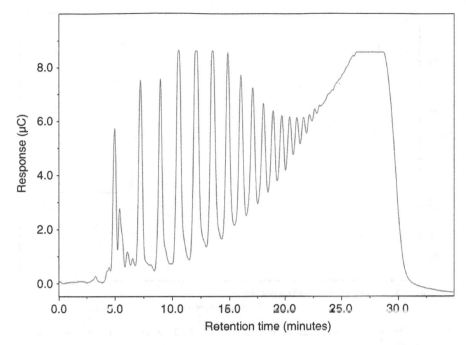

Fig. 2.1 Chitosan oligosaccharides (DP 2–16) separated by HILIC

TOF MS. In the MALDI-TOF MS, chitosan oligomer contained quasi-molecular ions [M+Na]$^+$ and [M+K]$^+$. Under their conditions, COS up to 12mers can be purified [1]. Li et al. also characterized the structures of degraded chitosan by MADLDI-TOF MS. In their opinion, chitosan oligomers exhibited similar signal strengths, irrespective of structure, when examined by MALDI-TOF MS. Therefore, the spectra allow the relative quantities of constituents of COS mixture. The results showed that their products were composed mainly of COS with DP 3–8 [10].

The high sensitivity of the MS detection allows the off-line MS method be a simple and time-saving method. An off-line TOF-MS analysis of COS was previously reported by Wu et al. In their experiment, COS with DP up to 18 was separated and identified [11]. Nicole et al. also reported an off-line MS method; they analyzed COS with DP from 2 to 6 by real-time electrospray ionization-mass spectrometry [12]. Recently Kim developed an LC MS/MS method for analysis of COS with DP 2–12. In their method, an amine column (Shodex, Japan) was used for HPLC separation where gradient elution using acetonitrile and water TOF, combined with a quadruple ion trap analyzer. It is the first report of online LC MS/MS analysis of COS with DP higher than 6. The LC MS/MS method can give comprehensive information, especially in the quantitative aspect [13].

HILIC uses hydrophilic stationary phases with reversed-phase-type eluents. It can separate and enrich polar compounds. COS with many polar groups can offer

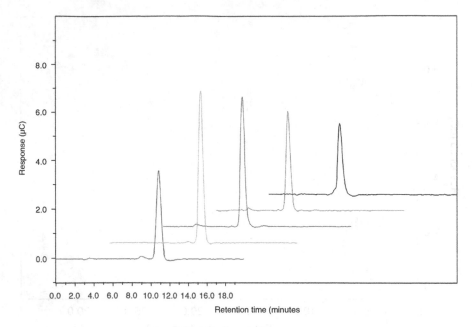

Fig. 2.2 Chitosan oligosaccharide monomers (DP 2–6) separated by cation-exchange chromatography

sufficient hydrophilic interactions with polar stationary phases. This method can also be applied for the separation of other various kinds of carbohydrates, ranging from carrageenan oligosaccharides and alginate oligosaccharides (AOS) to higher molecular weight fructooligosaccharides (DP 2–50) [14]. Purification of these oligosaccharides can be carried out by this strategy on a semipreparative scale (Fig. 2.2).

We also developed a cation-exchange column separation method for separating and purifying COS monomers (DP 2–6); due to that it was positive charged. Xiong et al. separated five fractions from dimer to hexamer by a linear gradient solution of HCl on a cation-exchange resin. Each purified fraction was analyzed by HPLC using PAD on a CarboPac PA1 anion-exchange column (250×4 mm I.D. with a particle size of 10 μm, Dionex Corp., USA), with linear gradient elution of 0–0.1 mol/L NaOH at a flow rate of 1 ml/min and at a column temperature of 30 °C [15]. Cation-exchange column separation method has the advantages of low cost, simple operation, and convenience; the technical line is mature, the used filling materials can be reused, and the method can be used for large-scale production. Besides COS separation, this method was also applied for xylooligosaccharide separation successfully.

Li et al. also developed an ion-exchange column chromatography method for COS separation. They separated COS with DP>6 by CM sepharose fast flow column (2.6 cm×50 cm) according to their numbers of amino group. Five fractions with DP>6 were collected using successively increasing concentration of NaCl. The separation fractions were analyzed by HPLC, which mainly contained oligomers with DP 6–7 (41.31%, 50.22%), DP 7–8 (22.47%, 70.13%), DP 9–10 (53.06%, 27.99%), DP 10–12 (18.45%, 49.36%, 22.31%), and DP>12, respectively [16].

Wei et al. isolated and purified chitosan pentamer and chitosan hexamer from chitosan oligosaccharide mixture by ultra-filtration, nano-filtration, ethanol precipitation, and the CM-Sephadex C-25 column methods. HPLC analysis was performed on a Waters 2690 liquid chromatography (Waters Corp., USA) with a differential RI detector. COS was separated on a Shodex Asahipak NH2P-50 4E column (4.6 mm × 250 mm, Shodex, Japan). The optimum separation of HPLC was carried out with a mobile phase consisting of acetonitrile/water (70/30, v/v) at a flow rate of 1.0 ml/min. The column temperature was set at 30 °C. Results showed that chitosan hexamer was separated successfully which consisted of (GLcN)6 (93.11%) with little (GLcN)5 (6.89%). But chitosan pentamer was still mixed with (GlcN)4 (59.84%) and (GLcN)5 (40.16%) [17].

Franta Le et al. reported an immobilized metal affinity chromatography method for chitosan oligomers separation, based on the differences in the interactions of chelated copper (II) ions with various oligomers (dimers, trimers, tetramers). The DP of COS plays an important role in the interaction between Cu^{2+} and oligomers. The retention capacities of COS of the columns were between 2 and 6 mg/cm^3 depending on the chelating group of the columns. In their method, COS was separated and/or enriched up to 95% for dimer and trimer and 90% for the tetramer, with yields of 60–95% [18].

Chitin Oligosaccharides

Chitin oligosaccharide is produced by hydrolysis of chitin that has been refined from crab and shrimp shells. Structurally, it is an oligosaccharide that takes the form of several N-acetylglucosamine (GlcNAc) molecules linked together. It is a well-known elicitor that has been known to induce various defense responses in a wide range of plant cells including both monocots and dicots. Recently several proteins involved in the perception and transduction of chitin oligosaccharide elicitor were characterized to further clarify the molecular mechanism. Here we summarized several analysis and separation methods of chitin oligosaccharides.

UV detector is the most common liquid chromatography detector, which can be used to analyze chitin oligosaccharides. Krokeide et al. separated mixtures of (GlcNAc)$_4$ and (GlcNAc)$_2$ by normal-phase HPLC using a Tosoh TSK amide 80 column (0.2 × 25 cm, Tosoh Corp., Japan) with an amide 80 guard column. The sample size was 50 μL and the chitin oligosaccharides were eluted isocratically at 0.25 ml/min with 70% (v/v) acetonitrile at room temperature. The chitin oligosaccharides were monitored by measuring absorbance at 210 nm and the (GlcNAc)$_4$ concentrations were quantified by measuring peak areas and by comparing these to those of standard samples with known concentrations of (GlcNAc)$_4$. Using these

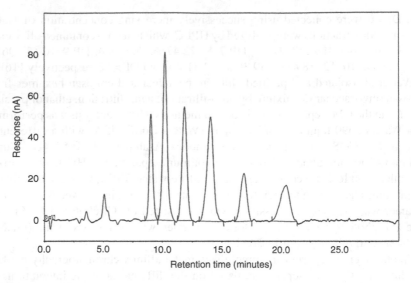

Fig. 2.3 Chitin oligosaccharide monomers (DP 2–6) separated by using Asahipak NH$_2$P-50 4E

standard samples, it was established that there was a linear correlation between the peak area and the analyzed (GlcNAc)$_4$ concentration within the concentration range 0.5–300 M used in this study [19].

In our lab chitin oligosaccharides can be analyzed by an amide column combined with PAD detector or CAD detector. A method using Shodex Asahipak NH$_2$P-50 4E column (Shodex, Japan) coupled to PAD detector was developed. The elution was performed with 70 % acetonitrile at a flow rate of 0.5 ml/min at 35 °C. The chitin oligomers of DP 2–6 can be well separated under such a condition (Fig. 2.3).

TSK-Gel Amide-80 column (Tosoh Corp., Japan) and CAD detector were also suitable for chitin oligosaccharide separation; 70 % acetonitrile at a flow rate of 0.7 ml/min at 25 °C was chosen as the elution condition. With the TSK-Gel Amide-80 column, it is possible to see double peaks due to separation of the α- and β-anomers of reducing sugars [20] (Fig. 2.4).

Combining HPLC and electrospray MS, Suginta et al. analyzed enzymatic properties of wild-type and active site mutants of chitinase A from *Vibrio carchariae*. Their approach allowed the separation of α- and β-anomers and the simultaneous monitoring of chitin oligosaccharide products down to picomole levels [21]. Tang reported a HILIC-MS method to evaluate the deacetylase activity of acetyl xylan esterase variants by profiling the deacetylated chitooligosaccharide products. The HILIC-MS/MS sequencing revealed that 30 different deacetylation products ranging from (GlcNAc)5(GlcN)1 to (GlcNAc)1(GlcN)5 and isomers thereof were produced. Coupling HPLC with MS (LC–MS) can provide two-dimensional separations; highly similar chitin oligosaccharides can be resolved based on retention time and mass signature, giving information on the precise identification of peaks unless pure standards are available [22].

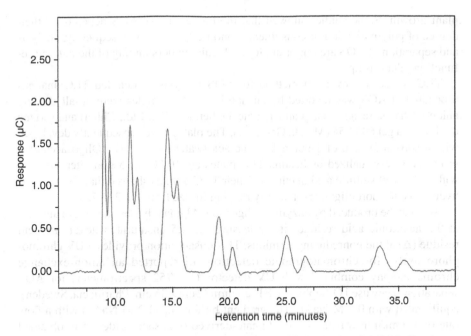

Fig. 2.4 Chitin oligosaccharide monomers (DP 2–6) separated by using TSK-Gel Amide-80 column

Lopatin et al. reported a semipreparative-scale chromatography of (GlcNAc)$_{2-7}$ on a reversed-phase C-16 HPLC column. The GlcNAcs was separated on a column with Sephadex G-25 sf (3×70 cm, Pharmacia, Sweden) in water at a flow of 225 ml/h. Portions of 300 mg of hydrolyzate in 1 ml of water were applied on the column. Fractions of 10 ml each were concentrated and analyzed by means of reversed-phase chromatography on a LiChrospher 100RP-18 column (4×250 mm, Merck, Germany) with a variable wavelength detector UV-VIS (Knauer, Germany). The fractions collected after the above gel filtration column having identical composition according to RP-HPLC were combined. Then a semipreparative purification of individual GlcNAcs by means of reversed-phase chromatography was performed with the same chromatographic equipment [23]. Aiba used gel filtration chromatography to separate (GlcNAc)n ($n=2$–5). The yields of four GlcNAcs were 16.3, 10.3, 18.0, and 1.7 mg respectively, from the hydrolyzates of 72 % N-acetylated chitosan (100 mg) [24].

Alginate Oligosaccharides

Alginate oligosaccharide (AOS) is a kind of acidic oligosaccharide, composed of α-L-guluronate and β-D-mannuronate, linking with 1,4-O-glycosidic bonds. AOS have been attracting considerable attention due to their bioactivity of promoting plant growth. Some studies showed that the bioactivity of AOS depends on their degree of polymerization or constituent monomer [25, 26]. Consequently, analysis and separation of AOS are important for molecular understanding of their structure-function relationship.

TLC is also a convenient method for AOS analysis. A detailed TLC analysis procedure of AOS was reported by our lab [27]. The samples were desalted using anion-exchange resins (Lianguan Biological Chemical Co., Ltd., China) and spotted on the silica gel 60 F254 (Merck, Germany). The plates were subsequently developed with a solvent system of 1-butanol/formic acid/water (4:6:1, v/v). Oligosaccharide products were visualized by heating TLC plates at 110 °C for 5 min after spraying with 10 % (v/v) sulfuric acid in ethanol. Their TLC profiles showed that alginate was hydrolyzed to short oligomers, majority ranging in size from DP 2 to 7.

AOS can be produced by enzymatic digestion. Alginate lyases can cleave alginate at the hexuronic acid residue sites releasing the 4,5-unsaturated hexuronic acid residue (Δ) at the nonreducing terminus. This unsaturation provides a UV chromophore useful for chromatographic detection. Li reported an anion-exchange chromatography combined with UV detector (UV-752 spectrometer) for AOS separation. They used Q-Sepharose F.F. column (2 cm × 35 cm, Pharmacia, Sweden), equilibrated with 0.2 M NaAc. The gradient buffer (0.2–1.2 M NaAc) with a flow rate of 1.0 ml/min was used. Six alginate-derived oligosaccharides were obtained and separated by this method [27].

Fu et al. developed a HILIC with UV detector for AOS analysis. An analytical "click" maltose (100 × 4.6 mm id) column was used at a flow rate of 1.0 ml/min and a temperature of 30 °C. The mobile phase consisted of acetonitrile (A) and ammonium formate solution (100 mM, pH 3.0, B). The gradient is as follows: 0–30 min: linear from 30 % B to 50 % B, 30–40 min: hold 50 % B. Under the eluent conditions, the separation of AOS was achieved and their DPs ranged from 2 to 6 [14].

Ballance et al. applied high-performance anion-exchange chromatography (HPAEC) with PAD to study AOS distributions. The chromatography system and conditions used are described below: IonPac AS4A (4 mm × 250 mm) anion-exchange column connected to an IonPac AG4A (4 mm × 50 mm) guard column was used. Chromatography was performed at a flow rate of 1 ml/min. Buffer A was 0.1 M NaOH, prepared from a carbonate-free 50 % (w/w) NaOH solution, and buffer B was 1 M NaAc in 0.1 M NaOH. Linear gradients of acetate were produced to elute the samples by increasing the concentration of buffer B from 0 to 70 % over 80 min. Column effluent was monitored with a PAD on an Au working electrode and Ag/AgCl reference electrode. The sequence of potentials applied to the electrode was as follows: E1 = 0.05 V (480 ms, integrating from 280 to 480 ms), E2 = 0.6 (120 ms), and E3 = −0.8 V (300 ms) at a sensitivity of 100 nC. They obtained a high-resolution separation of oligosaccharides according to chain length until they exceeded between 30 and 35 monomer units. They also found that different monomer sequences have different retention times on the anion-exchange resin. Mannuronan oligosaccharides were more strongly retained than those of the same chain length derived from G-blocks. Their results showed that HPAEC–PAD is a useful tool to determine the sugar residue distribution of AOS [28].

Zhang et al. developed a gel filtration method for fractionation of AOS from either lyase or acid hydrolysis by Bio-Gel P6 column. The AOS mixture (2 ml, 100 mg/ml) was loaded onto a Bio-Gel P6 column (2.6×100 cm, Bio-Rad, USA) and eluted by 0.5 M NH_4HCO_3 at a flow rate of 15 ml/h at room temperature. The sample was monitored online by an RI detector (Gilson 132). They rechromatographed oligosaccharide fractions on the same column and obtained pure homooligosaccharide fractions, oligomannuronic acids (DP 2–5), and oligoguluronic acids (DP 2–5). The fractions of oligosaccharides derived from lyase digestion were further fractionated by preparative SAX-HPLC (ÄKTA™ FPLC, GE Healthcare Corp., USA). The column was Spherisorb S5 SAX (20×250 mm, GE Healthcare Corp., USA), and was eluted with a linear gradient from 0 to 0.4 mol/L NaCl solution at a flow rate of 2 ml/min for 2.5 h, and seven unsaturated oligosaccharides, $\triangle$G, $\triangle$GG, $\triangle$MG, $\triangle$GGG, $\triangle$MGG, $\triangle$GMG, and $\triangle$MMM were separated and identified [29].

MS analysis is an effective method for AOS analysis. Huang et al. profiled the AOS composition by electrospray ionization mass spectrometry (ESI–MS) [30]; samples were loaded onto a microcrystalline cellulose column followed by a cation-exchange column to remove proteins and salts, and then concentrated, dried, and dissolved in 1 ml of methanol. After centrifugation, the supernatant was loop-injected to an LTQ XL linear ion trap mass spectrometer (Thermo Fisher Scientific, USA). The oligosaccharides were detected in a positive-ion mode using the following settings: ion source voltage, 4.5 kV; capillary temperature, 275–300 °C; tube lens, 250 V; sheath gas, 30 arbitrary units (AU); scanning the mass range, 150–2000 m/z. Ion peaks represented the mono-dehydrated sodium adducts of alginate oligosaccharides from DP2 to DP10.

As we have known, AOS is composed of mannuronic acid (M) and guluronic acid (G); the composition (M/G ratio) and sequence determination of the hexuronic acid M and G residues are very important for better understanding of the structure-function relationship. MS can be used for sequence analysis of AOS. Zhang et al. reported a sequence analysis method of alginate-derived oligosaccharides by negative-ion electrospray ionization tandem mass spectrometry (ESI-MS/MS) with collision-induced dissociation (CID). Unique fragmentation can arise at certain monosaccharide residues with specific linkages under CID-MS/MS conditions. By using this information on sequence and linkages, they successfully obtained and analyzed sequence of 16 homo- and hetero-oligomeric fragments [29].

Oligogalacturonic Acid

Oligogalacturonic acid (OGA) is a linear 1,4-linked alpha-D-galacturonic acid oligomers. Plant cell wall is a source of OGA fragments generated during plant-pathogen interactions through the enzymatic hydrolysis of polysaccharides. Thus OGAs can act as signaling molecules to initiate cascades of events that result in promoting plant growth or inducing plant defense. A specific degree of polymerization range of OGAs has been associated with optimal biological activity in some of the biological processes [31, 32]. Thus it is important to investigate the DP distribution of OGAs and further explore its activity.

OGA also could be analyzed by RI detector as Stoll et al. reported [33]. They used a Cyclobond I 2000 column (250×4.6 mm, Sigma-Aldrich, USA) and a volatile mobile phase consisting of ammonium formate and methanol. The separation was operated at a temperature of 40 °C. The elution was performed with two different isocratic systems: (1) ammonium formate (55 mM, pH 4):methanol (70:30, v/v) at a flow rate of 1 ml/min, and (2) ammonium formate (100 mM, pH 4):methanol (70:30, v/v) at 0.8 ml/min. Under the procedure described above, OGA up to DP of 7 can be baseline separated within 40 min [33].

HPAEC-PAD is used extensively to determinate the DP distribution of OGA, usually using the CarboPac PA column to separate. Camejo reported a chromatographic method to determine the DP of OGAs. HPAEC was performed with a CarboPac PA100 column and a PAD from Dionex. The mobile phases were degassed with helium in order to prevent the absorption of carbon dioxide and the transformation to carbonate. The analysis was carried out at 1 ml/min with a linear gradient of 200–500 mM sodium acetate in 100 mM sodium hydroxide (0–50 min) and 500–800 mM sodium acetate in 100 mM sodium hydroxide (50–65 min). The column was reconditioned by washing with 800 mM sodium acetate containing 100 mM sodium hydroxide and then re-equilibrated with the starting buffer solution. They chose OGAs with a DP of 7–15 for further research. The effects of OGAs on root length, extracellular alkalinization, and O^{2-} accumulation in alfalfa were investigated. They found that OGAs could promote intact root alfalfa growth at 25, 50, and 75 µg/ml concentrations, and it is not related to extracellular alkalinization [34]. HPAEC-PAD has proven to be an effective method for OGAs' preparative-scale separation. Hotchkiss Jr et al. detected OGAs up to DP 20 by using preparative HPAEC-PAD. Using a CarboPac PA1 column (21×250 mm, Dionex Corp., USA) and a 5 ml/min nonlinear potassium acetate (pH 7.5) gradient in a 110-min run, they separated milligram quantities of OGAs enriched in DP 10–20 [35]. This method was simple and faster, had higher sample loading capacity, and allowed for the isolation of higher DP oligogalacturonic acids.

MS has been successfully applied to the characterization of OGA structure. Zhu et al. systematically investigated the behavior of OGA under different conditions using ESI-MS [36]. They considered three major problems that influenced analysis of OGAs, including cation adduction, fragmentation, and non-covalent binding. An optimized condition was 2 M acetic acid and 2.5 % TEA in 80 % acetonitrile as modifiers, and capillary voltage from 11 to 44 V. Under such condition, a clean and clear spectrum of trigalacturonic acid can be obtained. Doner et al. isolated individual OGA by step-gradient elution (sodium formate, pH 4.7) from the macroporous strong base anion-exchange resin AG MP-1 in the formate form. Pure oligomers to heptagalacturonic acid were isolated in a single run, including gram quantities of tri-, tetra-, and pentagalacturonic acid. The individual OGAs were characterized by fast-atom bombardment mass spectrometry in positive and negative modes [37]. Nowadays, the analysis of OGAs could be easily realized by various different methods;

however, the separation and characterization of pectic oligosaccharides still remain challenging due to their complex mixtures. Recently, Leijdekkers developed a new method for pectic oligosaccharide analysis using HILIC coupled to traveling-wave ion mobility mass spectrometry. This method enabled the simultaneous separation and characterization of complex mixtures of various isomeric pectic oligosaccharides. In this study, some novel structure features were identified for the first time: glucuronic acid was attached to O-3 or to O-2 of galacturonic acid residues and a single galacturonic acid residue within an oligomer could contain both an acetyl group and a glucuronic acid substituent [38].

Carrageenan Oligosaccharides

Carrageenans are water-soluble galactans extracted from some red seaweeds. They have a repeating backbone of 3-linked b-D-galactose (G-units) and 4-linked a-D-galactose (D-units), containing a varying degree of sulfate. Naturally derived carrageenan seldom has a regularly repeating structure and is more complicated. Usually there are three main types of carrageenan, which differ in their degree of sulfation. *Kappa*-carrageenan has one sulfate group per disaccharide. *Iota*-carrageenan has two sulfates per disaccharide and *Lambda*-carrageenan has three sulfates per disaccharide.

Carrageenan oligosaccharide has been reported to promote plant growth and elicit plant defense responses. The biological activity of carrageenan oligosaccharide is considered to be related to the number and position of sulfate groups in its structure [39, 40]. In order to further investigate the structure and activity relationship of carrageenan oligosaccharide, effective chromatographic analysis and separation method becomes very necessary.

Knutsen et al. developed a rapid size-exclusion chromatography (SEC) method with RI detector for the separation and analysis of carrageenan oligosaccharides. In their work, Superdex 30 columns were used both for analytical and preparative applications. *Kappa*- and *iota*-carrageenan oligosaccharides of DP from 2 to 12 could be baseline separated in a 20-min run on an analytical scale. In the semi-preparative experiment, oligosaccharides at an mg scale were obtained in one single run only taking about 1 h. The system is also well suited to separate oligosaccharides derived from alginic acid [41].

The work by Lafosse's group suggests that different chromatographic conditions are required for analysis of *iota*-carrageenan oligomers and *kappa*-carrageenan oligomer due to the number of sulfate groups. They isolated *kappa*-carrageenan oligomers on a Spherisorb ODS1 column (250×4 mm, Waters Corp., USA) using ion-pair liquid chromatography coupled with ELSD. Heptylamine (5 mM, pH 4) was chosen as the ion-pairing agent and MeOH as the organic modifier in a gradient mode. This analytical method proved to be able to separate the oligomers of *kappa*-carrageenan up to 16 [42]. They also presented several separation methods for oligosaccharides of *iota*-carrageenan with ion exchangers, PGC and C_{18} columns, all coupled with ELSD. The oligomers were then isolated and characterized off-line with ESI-MS in the positive-ion mode. Compared to oligo-*kappa*-carrageenans, the additional sulfate group in the unit of *iota*-carrageenans significantly affected the separation mechanisms on ion-exchange chromatography, porous graphitic carbon (PGC), and ion-pair chromatography. Both with these separation methods, they could not detect any peak further than the octasaccharide [43].

Subsequently, this group developed an online LC/ESI-MS method for *kappa*-carrageenan oligosaccharide analysis with PGC column. PGC column presents simultaneous hydrophobic and electronic interactions that permit better elution control for *kappa*-carrageenan oligosaccharides. Meanwhile ESI-MS can detect the oligosaccharides as fully deprotonated form, by directly detecting the stable deprotonated sulfated ions. The position of the sulfate groups could be determined [44].

Aguilan et al. successfully analyzed the structure of *kappa*-carrageenan oligosaccharide with DP up to 10 by positive-mode nano-ESI-FTICR-MS together with MS/MS using sustained off-resonance irradiation-collision-induced dissociation (SORI-CID). Through observing the B- and Y-types of fragmentation obtained by glycosidic bond cleavage reactions, complete sequencing of the oligosaccharide samples can be determined. Meantime the positions of the labile sulfate substituents were observable using SORI-CID, enabling the determination of the sequence of the sulfated residues. The high mass accuracy and sensitivity of the method proved to be very effective for the analysis of carrageenan oligosaccharides. This technique can also be applied to the analysis of other carbohydrates that are difficult to analyze using other MS methods [45].

Fatema et al. reported the behavior of *kappa*- and *iota*-carrageenans in MALDI-TOF mass spectrometry. The MALDI prompt and post-source decay (PSD) fragmentation processes were studied. Their results suggested that prompt fragmentation observed in the MALDI-MS analysis of a mixture of kappa-carrageenan oligosaccharides was mainly due to glycosidic C-cleavages. *Iota*-carrageenan oligosaccharides with more than one sulfate group gave desulfation, but glycosidic fragmentation was more important [46]. Gonçalves et al. used two different ESI-mass spectrometric techniques (ESI-CID MS/MS and ESI MS^n) for carrageenan oligosaccharide analysis. They also found that the relative positioning of the sulfate groups and type of monosaccharide unit affect the rate of cleavage of the glycosidic bonds [47].

Conclusion

At present, ion-exchange chromatography coupled with electrochemical detection is still a powerful analytical technique for oligosaccharide routine monitoring or research application. It is almost successfully employed for all kinds of oligosaccharide analysis; however, the incompatibility with MS restricts its application in determination of oligosaccharide with complex structure. Hydrophilic interaction liquid chromatography has recently been introduced as an efficient technique for oligosaccharide separation. It is very easy to couple with different detectors and suitable for separating most oligosaccharides. Especially, the compatibility of MS makes this technique a great impact on the analysis and separation of oligosaccharides. Polar oligosaccharides always show good on-column retention; it is suited to the sensitive LC-MS analysis, which overcomes the drawback encountered in ion-exchange chromatography. The spectrometric methods that have been reviewed in this chapter include analysis method for basic oligosaccharides, neutral oligosaccharides, and acidic oligosaccharides. Considering that the basic principle of oligosaccharide separation and analysis is same, the methods included in this chapter could be referenced for general oligosaccharide analysis.

Although a large number of analytical techniques have been developed for oligo-saccharides, the efficient large-scale separation method is still scare, as well as the in-depth bioactivity description of oligosaccharides with definite structure is still an unsolved problem. Oligosaccharides participate as elicitors in the cell cell recognition process in plants. Analytical methods of oligosaccharide can be usefully applied to gain insight into the biochemistry of these biological processes. The most typical example of oligosaccharide monomer application is that chitin oligosaccharide monomers were successfully used for its receptor identification and characterization [48]. It is predictable that further structure-activity relationship exploration and mechanisms research by obtaining oligosaccharide with clear structure, high purity, and yield will become a very accessible perspective.

References

1. Cabrera JC, Van Cutsem P. Preparation of chitooligosaccharides with degree of polymerization higher than 6 by acid or enzymatic degradation of chitosan. Biochem Eng J. 2005;25(2):165–72.
2. Tanuma H, Saito T, Nishikawa K, Dong T, Yazawa K, Inoue Y. Preparation and characterization of PEG-cross-linked chitosan hydrogel films with controllable swelling and enzymatic degradation behavior. Carbohydr Polym. 2010;80(1):260–5.
3. Kittur FS, Kumar ABV, Varadaraj MC, Tharanathan RN. Chitooligosaccharides—preparation with the aid of pectinase isozyme from Aspergillus niger and their antibacterial activity. Carbohydr Res. 2005;340(6):1239–45.
4. Jeon YJ, Kim SK. Production of chitooligosaccharides using an ultrafiltration membrane reactor and their antibacterial activity. Carbohydr Polym. 2000;41(2):133–41.
5. Qin CQ, Du YM, Xiao L, Li Z, Gao XH. Enzymic preparation of water-soluble chitosan and their antitumor activity. Int J Biol Macromol. 2002;31(1–3):111–7.
6. Kumar ABV, Varadara MC, Gowda LR, Tharanathan RN. Characterization of chito-oligosaccharides prepared by chitosanolysis with the aid of papain and Pronase, and their bactericidal action against Bacillus cereus and Escherichia coli. Biochem J. 2005;391:167–75.
7. Hsiao YC, Lin YW, Su CK, Chiang BH. High degree polymerized chitooligosaccharides synthesis by chitosanase in the bulk aqueous system and reversed micellar microreactors. Process Biochem. 2008;43(1):76–82.
8. Dong XF, Shen AJ, Gou ZM, Chen DL, Liang XM. Hydrophilic interaction/weak cation-exchange mixed-mode chromatography for chitooligosaccharides separation. Carbohydr Res. 2012;361:195–9.

9. Tokuyasu K, Ono H, OhnishiKameyama M, Hayashi K, Mori Y. Deacetylation of chitin oligosaccharides of dp 2–4 by chitin deacetylase from Colletotrichum lindemuthianum. Carbohydr Res. 1997;303(3):353–8.

10. Li J, Du YM, Yang JH, Feng T, Li AH, Chen P. Preparation and characterisation of low molecular weight chitosan and chito-oligomers by a commercial enzyme. Polym Degrad Stabil. 2005;87(3):441–8.

11. Wu HG, Yao Z, Bal XF, Du YG, Lin BC. Anti-angiogenic activities of chitooligosaccharides. Carbohydr Polym. 2008;73(1):105–10.

12. Dennhart N, Fukamizo T, Brzezinski R, Lacombe-Harvey ME, Letzel T. Oligosaccharide hydrolysis by chitosanase enzymes monitored by real-time electrospray ionization-mass spectrometry. J Biotechnol. 2008;134(3–4):253–60.

13. Kim J, Kim J, Hong J, Lee S, Park S, Lee JH, et al. LC-MS/MS analysis of chitooligosaccharides. Carbohydr Res. 2013;372:23–9.

14. Fu Q, Liang T, Zhang XL, Du YG, Guo ZM, Liang XM. Carbohydrate separation by hydrophilic interaction liquid chromatography on a 'click' maltose column. Carbohydr Res. 2010;345(18):2690–7.

15. Xiong CN, Wu HG, Wei P, Pan M, Tuo YQ, Kusakabe I, et al. Potent angiogenic inhibition effects of deacetylated chitohexaose separated from chitooligosaccharides and its mechanism of action in vitro. Carbohydr Res. 2009;344(15):1975–83.

16. Li KC, Xing RE, Liu S, Qin YK, Li B, Wang XQ, et al. Separation and scavenging superoxide radical activity of chitooligomers with degree of polymerization 6-16. Int J Biol Macromol. 2012;51(5):826–30.

17. Wei XL, Wang YF, Xiao JB, Xia WS. Separation of chitooligosaccharides and the potent effects on gene expression of cell surface receptor CR3. Int J Biol Macromol. 2009;45(4):432–6.

18. Le Devedec F, Bazinet L, Furtos A, Venne K, Brunet S, Mateescu MA. Separation of chitosan oligomers by immobilized metal affinity chromatography. J Chromatogr A. 2008;1194(2):165–71.

19. Krokeide IM, Synstad B, Gaseidnes S, Horn SJ, Eijsink VGH, Sorlie M. Natural substrate assay for chitinases using high-performance liquid chromatography: a comparison with existing assays. Anal Biochem. 2007;363(1):128–34.

20. Wang XH, Zhao Y, Tan HD, Chi NY, Zhang QF, Du YG, et al. Characterisation of a chitinase from Pseudoalteromonas sp. DL-6, a marine psychrophilic bacterium. Int J Biol Macromol. 2014;70:455–62.

21. Suginta W, Vongsuwan A, Songsiriritthigul C, Svasti J, Prinz H. Enzymatic properties of wild-type and active site mutants of chitinase A from Vibrio carchariae, as revealed by HPLC-MS. FEBS J. 2005;272(13):3376–86.

22. Tang MC, Nisole A, Dupont C, Pelletier JN, Waldron KC. Chemical profiling of the deacetylase activity of acetyl xylan esterase A (AxeA) variants on chitooligosaccharides using hydrophilic interaction chromatography-mass spectrometry. J Biotechnol. 2011;155(2):257–65.

23. Lopatin SA, Ilyin MM, Pustobaev VN, Bezchetnikova ZA, Varlamov VP, Davankov VA. Mass-spectrometric analysis of N-acetylchitooligosaccharides prepared through enzymatic-hydrolysis of chitosan. Anal Biochem. 1995;227(2):285–8.

24. Aiba S. Preparation of N-acetylchitooligosaccharides from lysozymic hydrolysates of partially N-acetylated chitosans. Carbohydr Res. 1994;261(2):297–306.

25. Natsume M, Kamo Y, Hirayama M, Adachi T. Isolation and characterization of alginate-derived oligosaccharides with root growth-promoting activities. Carbohydr Res. 1994;258:187–97.

26. Iwasaki K, Matsubara Y. Purification of alginate oligosaccharides with root growth-promoting activity toward lettuce. Biosci Biotechnol Biochem. 2000;64(5):1067–70.

27. Li LY, Jiang XL, Guan HS, Wang P. Preparation, purification and characterization of alginate oligosaccharides degraded by alginate lyase from Pseudomonas sp. HZJ 216. Carbohydr Res. 2011;346(6):794–800.

28. Campa C, Oust A, Skjak-Braek G, Paulsen BS, Paoletti S, Christensen BE, et al. Determination of average degree of polymerisation and distribution of oligosaccharides in a partially acid-hydrolysed homopolysaccharide: a comparison of four experimental methods applied to mannuronan. J Chromatogr A. 2004;1026(1–2):271–81.

29. Zhang ZQ, Yu GL, Zhao X, Liu HY, Guan HS, Lawson AK, et al. Sequence analysis of alginate-derived oligosaccharides by negative-ion electrospray tandem mass spectrometry. J Am Soc Mass Spectrom. 2006;17(7):1039.

30. Huang LS, Zhou JG, Li X, Peng Q, Lu H, Du YG. Characterization of a new alginate lyase from newly isolated Flavobacterium sp. S20. J Ind Microbiol Biotechnol. 2013;40(1):113–22.

31. Navazio L, Moscatiello R, Bellincampi D, Baldan B, Meggio F, Brini M, et al. The role of calcium in oligogalacturonide-activated signalling in soybean cells. Planta. 2002;215(4):596–605.

32. Simpson SD, Ashford DA, Harvey DJ, Bowles DJ. Short chain oligogalacturonides induce ethylene production and expression of the gene encoding aminocyclopropane 1-carboxylic acid oxidase in tomato plants. Glycobiology. 1998;8(6):579–83.

33. Stoll T, Schieber A, Carle R. High-performance liquid chromatographic separation and on-line mass spectrometric detection of saturated and unsaturated oligogalacturonic acids. Carbohydr Res. 2002;337(24):2481–6.

34. Camejo D, Marti MC, Jimenez A, Cabrera JC, Olmos E, Sevilla F. Effect of oligogalacturonides on root length, extracellular alkalinization and O-2(–)-accumulation in alfalfa. J Plant Physiol. 2011;168(6):566–75.

35. Hotchkiss AT, Leerinier SL, Hicks KB. Isolation of oligogalacturonic acids up to DP 20 by preparative high-performance anion-exchange chromatography and pulsed amperometric detection. Carbohydr Res. 2001;334(2):135–40.

36. Zhu L, Lee HK. Preliminary study of the analysis of oligogalacturonic acids by electrospray ionization mass spectrometry. Rapid Commun Mass Spectrom. 2001;15(12):975–8.

37. Doner LW, Irwin PL, Kurantz MJ. Preparative chromatography of oligogalacturonic acids. J Chromatogr. 1988;449(1):229–39.

38. Leijdekkers AGM, Huang JH, Bakx EJ, Gruppen H, Schols HA. Identification of novel isomeric pectic oligosaccharides using hydrophilic interaction chromatography coupled to traveling-wave ion mobility mass spectrometry. Carbohydr Res. 2015;404:1–8.

39. Hashmi N, Khan MMA, Moinuddin, Idrees M, Khan ZH, Ali A, et al. Depolymerized carrageenan ameliorates growth, physiological attributes, essential oil yield and active constituents of Foeniculum vulgare Mill. Carbohydr Polym. 2012;90(1):407–12.

40. Lemonnier-Le Penhuizic C, Chatelet C, Kloareg B, Potin P. Carrageenan oligosaccharides enhance stress-induced microspore embryogenesis in Brassica oleracea var. italica. Plant Sci. 2001;160(6):1211–20.

41. Knutsen SH, Sletmoen M, Kristensen T, Barbeyron T, Kloareg B, Potin P. A rapid method for the separation and analysis of carrageenan oligosaccharides released by iota- and kappa-carrageenase. Carbohydr Res. 2001;331(1):101–6.

42. Antonopoulos A, Favetta P, Helbert W, Lafosse M. Isolation of kappa-carrageenan oligosaccharides using ion-pair liquid chromatography—characterisation by electrospray ionisation mass spectrometry in positive-ion mode. Carbohydr Res. 2004;339(7):1301–9.

43. Antonopoulos A, Favetta P, Lafosse M, Helbert W. Characterisation of iota-carrageenans oligosaccharides with high-performance liquid chromatography coupled with evaporative light scattering detection. J Chromatogr A. 2004;1059(1–2):83–7.

44. Antonopoulos A, Favetta P, Helbert W, Lafosse M. On-line liquid chromatography-electrospray ionisation mass spectrometry for kappa-carrageenan oligosaccharides with a porous graphitic carbon column. J Chromatogr A. 2007;1147(1):37–41.

45. Aguilan JT, Dayrit FM, Zhang JH, Ninonuevo MR, Lebrilla CB. Structural analysis of alpha-carrageenan sulfated oligosaccharides by positive mode nano-ESI-FTICR-MS and MS/MS by SORI-CID. J Am Soc Mass Spectrom. 2006;17(1):96–103.

46. Fatema MK, Nonami H, Ducatti DRB, Goncalves AG, Duarte MER, Noseda MD, et al. Matrix-assisted laser desorption/ionization time-of-flight (MALDI-TOF) mass spectrometry analysis of oligosaccharides and oligosaccharide alditols obtained by hydrolysis of agaroses and carrageenans, two important types of red seaweed polysaccharides. Carbohydr Res. 2010;345(2):275–83.
47. Goncalves AG, Ducatti DRB, Grindley TB, Duarte MER, Noseda MD. ESI-MS differential fragmentation of positional isomers of sulfated oligosaccharides derived from carrageenans and agarans. J Am Soc Mass Spectrom. 2010;21(8):1404–16.
48. Shibuya N, Kaku H, Kuchitsu K, Maliarik MJ. Identification of a novel high-affinity binding-site for N-acetylchittooligosaccharide elicitor in the membrane-fraction from suspension-cultured rice cells. FEBS Lett. 1994;348(1):107–8.

Chapter 3
Oligosaccharin Receptors in Plant Immunity

Tomonori Shinya, Yoshitake Desaki, and Naoto Shibuya

Abstract It has been well known that oligosaccharides generated from fungal, bacterial, or plant cell walls can induce various defense responses in plant cells. Traditionally they have been called as "elicitors" or "general elicitors" that induce defense responses in a wide range of plant species. These molecules are now classified as representative MAMPs/PAMPs (microbe/pathogen-associated molecular patterns) or DAMPs (damage/danger-associated molecular patterns) and believed to play important roles in plant immunity, as a trigger of the so-called pattern-triggered immunity (PTI). Biological activities of these oligosaccharide elicitors, including structure/function relationships, corresponding receptors, and downstream signaling, have been the subject of intense studies, by which these oligosaccharides became most well-characterized oligosaccharins. In this chapter, present knowledge on the plant receptors of carbohydrate elicitors generated from the cell walls of pathogenic microbes and also host plants is summarized. Mechanism of the activation of downstream signaling is discussed for chitin receptor, as this system has become the most well-characterized oligosaccharin receptor in plant immunity. It is also discussed how pathogenic microbes try to escape from the MAMP-mediated detection of their invasion by host plants.

Keywords MAMPs • PAMPs • DAMPs • Elicitor • Chitin • Peptidoglycan • β-glucan • Oligogalacturonides • Plant immunity • Receptor

T. Shinya, Ph.D.
Department of Life Sciences, School of Agriculture, Meiji University,
1-1-1 Higashi-Mita, Tama-ku, Kawasaki, Kanagawa 214-8571, Japan

Institute of Plant Science and Resources, Okayama University,
2-20-1, Chuo, Kurashiki, Okayama 710-0046, Japan
e-mail: shinyat@rib.okayama-u.ac.jp

Y. Desaki, Ph.D. • N. Shibuya, Ph.D. (✉)
Department of Life Sciences, School of Agriculture, Meiji University,
1-1-1 Higashi-Mita, Tama-ku, Kawasaki, Kanagawa 214-8571, Japan
e-mail: desaki@meiji.ac.jp; shibuya@meiji.ac.jp

© Springer Science+Business Media New York 2016   29
H. Yin, Y. Du (eds.), *Research Progress in Oligosaccharins*,
DOI 10.1007/978-1-4939-3518-5_3

Abbreviations

MAMP   Microbe-associated molecular pattern
PAMP   Pathogen-associated molecular pattern
DAMP   Damage/danger-associated molecular pattern
GPI   Glycosylphosphatidylinositol
LRR   Leucine-rich repeat
OGs   Oligogalacturonides
PGase   Polygalacturonase
PGN   Peptidoglycan
PRR   Pattern recognition receptor
PTI   Pattern-triggered immunity
RLK   Receptor-like kinase

Chitin Receptor

Chitin (β-1,4-linked polymer of N-acetylglucosamine) is a common component of fungal cell walls and its fragments, N-acetylchitooligosaccharides, have been shown to act as a potent MAMP elicitor in various plant systems [1, 2]. It has been expected that N-acetylchitooligosaccharides are generated during the infection process of fungal pathogens through the action of plant chitinases. The released oligosaccharides are perceived by the host plant to initiate various defense responses. The requirements for the size of chitin fragments are different depending on the experimental systems probably reflecting the specificity of corresponding receptors. In general, however, the elicitor activity increases with the increase of degree of polymerization up to the octasaccharide [3–5]. The sensing system for the chitin fragments seems to be very sensitive, sometimes responding to nano-molar or even lower concentrations of these oligosaccharides.

A series of biochemical studies indicated the presence of receptor molecules for chitin oligosaccharides in the plasma membrane of several plants [6–11]. The binding protein for chitin oligosaccharides was eventually purified from the plasma membrane of rice and named CEBiP [12]. Knockdown experiments of *CEBiP* gene indicated that CEBiP is a major chitin-binding protein on the cell surface and required for signaling, showing that CEBiP is a functional receptor. CEBiP homologues were also identified in *Arabidopsis*, *Medicago truncatula*, and barley [13–15]. CEBiP and its homologues were predicted to have two or three extracellular LysM motifs and a transmembrane domain or GPI-anchor but lack the intracellular domain expected for signal transduction. This observation suggested that some additional components are required for signaling through the plasma membrane. A receptor-like kinase (RLK) of *Arabidopsis*, CERK1 (chitin elicitor receptor kinase1), was identified as a molecule essential for chitin elicitor signaling by reverse genetic approach [16, 17]. CERK1 is a plasma membrane protein containing

three extracellular LysM motifs, a transmembrane domain, and an intracellular Ser/
Thr kinase domain. The knockout mutant of *CERK1* completely lost the ability to
respond to chitin elicitor and also exhibited decreased resistance to several fungal
pathogens [16, 17]. A functional homologue of CERK1 in rice, OsCERK1, was
also identified and shown to contribute to chitin signaling through the formation
of a receptor complex with CEBiP in the presence of chitin oligosaccharides [18].
The fact that CEBiP binds chitin oligosaccharides with a high affinity whereas
OsCERK1 does not bind chitin [12, 13] suggests that chitin oligosaccharides first
bind to CEBiP and the following ligand-induced formation of the CEBiP-OsCERK1
complex triggers the activation of OsCERK1 and downstream signaling.

Characterization of *Arabidopsis* chitin receptor system indicated the presence of
significant difference between rice and *Arabidopsis* receptor systems. In addition to
CERK1, which is essential for chitin perception [16], *Arabidopsis* possesses three
CEBiP homologs, LYM1-3, which contain extracellular LysMs and GPI-anchor.
Among *Arabidopsis* CEBiP homologs, AtCEBiP/LYM2 showed a high affinity
binding for chitin oligosaccharides similar to rice CEBiP [13]. In addition, AtCEBiP
was detected as a major chitin-binding protein in the microsomal membrane preparation of *Arabidopsis*. On the contrary to the expectation that AtCEBiP serves as a
component of *Arabidopsis* chitin receptor complex similar to rice, analysis of the
atcebip mutant and also the triple mutant for all CEBiP homologs indicated that
AtCEBiP does not contribute to chitin signaling. Wan et al. also reported similar
observation on chitin responses of the *atcebip* mutant [19]. On the other hands,
different characteristics of CERK1 were also reported for rice and *Arabidopsis*.
The fact that OsCERK1 does not bind chitin whereas CERK1 directly binds chitin
[13, 20] indicated that the receptor kinase CERK1 serves both for perception and
transduction of chitin oligosaccharides in *Arabidopsis*. These results indicated that
CERK1 is sufficient for chitin signaling by itself whereas rice requires a multicom-
ponent receptor consisting of CEBiP and OsCERK1. If plants acquired two types of
chitin receptor systems during evolution, which system is more general? Although
it is difficult to answer the question, the fact that CEBiP-like chitin-binding proteins
were detected in the plasma membranes from various plants [11] may indicate the
more general presence of the complex-type receptor system.

As described above, *Arabidopsis* AtCEBiP/LYM2 is biochemically very similar
to rice CEBiP but does not contribute to CERK1-mediated chitin signaling [13].
Is it a useless molecule left behind the evolution? Interestingly, *lym2* mutant showed
an increased susceptibility to pathogens as similar to *cerk1* mutant [21, 22]. Faulkner
et al. recently reported that LYM2 contributes to the regulation of molecular flux
via plasmodesmata and contributes to disease resistance against a necrotic fungus,
Botrytis cinerea, independently of CERK1 [21]. These studies clearly indicated
the presence of LYM2-dependent but CERK1-independent disease resistance
against fungal pathogens in *Arabidopsis*. It remains for future studies how the
LYM2-dependent resistance works.

In addition to the CEBiP and CERK1 homologs, several LysM-containing
proteins have been reported as a component of chitin receptor complex. LYK4, an
Arabidopsis LysM-RLK, was reported to play an important role in chitin signaling

in *Arabidopsis* [19]. Although the knockout mutant of CERK1 completely lost chitin responses, chitin-induced defense responses were partially observed in the knockout mutant of LYK4. The result indicated that LYK4 may assist chitin signaling in the receptor complex. Liu et al. reported that OsLYP4 and OsLYP6, rice LysM-containing proteins, are a dual-specificity receptor and involved in the perception of both chitin and peptidoglycan [23]. It remains to be clarified how and to what extent these LysM proteins contribute to chitin signaling in each plant species.

Concerning to the molecular mechanism of ligand-induced activation of chitin receptor, Liu et al. reported that chitin-induced dimerization of CERK1 is essential for chitin signaling [20]. They showed that the CERK1 dimerization requires longer chitin oligosaccharide such as chitooctamer, and a shorter oligosaccharide, $(GlcNAc)_5$, even inhibited $(GlcNAc)_8$-induced CERK1 dimerization and defense responses. Crystal structure of CERK1 ectodomain-chitopentamer complex indicated that CERK1 directly binds chitin oligosaccharides through LysM2, located at the center of the three LysMs in the extracellular domain [20]. On the other hand, epitope mapping by saturation transfer difference (STD)-NMR analysis as well as molecular modelling and docking studies indicated that rice CEBiP binds $(GlcNAc)_{7/8}$ in a unique sandwich-like manner, where two CEBiP molecules bind to one chitin oligosaccharide from the opposite surfaces through the central LysM region, resulting in the dimerization of CEBiP itself and also the formation of receptor complex consisting of both CEBiP and OsCERK1 [24].

Rapid dimerization and phosphorylation of CERK1 occurs within a few minutes after the treatment with chitin oligosaccharides [20, 25]. Phosphorylation of CERK1 seems to activate subsequent intracellular signaling pathway, such as the increase of $[Ca^{2+}]_{cyt}$, and activation of MAP kinases and Rho-type GTPase (RAC) [16, 19, 25, 26]. Several intracellular proteins have been reported to interact with CERK1/OsCERK1 and serve for the activation of downstream signaling. OsRacGEF1, a guanine nucleotide exchange factor for OsRac1, was identified as an OsRAC1 interactor and rapidly activated by chitin treatment [26]. After the chitin perception, OsCERK1 directly activates OsRacGEF1, resulting in the OsRac1-dependent defense responses. OsRLCK185 is a membrane-anchored cytoplasmic receptor-like kinase of rice, and directly interacts with OsCERK1 [27]. Following chitin perception, OsCERK1 phosphorylates OsRLCK185, which then seems to partially dissociate from OsCERK1 complex and activate downstream defense signaling. In *Arabidopsis*, PBL27 was identified as an ortholog of OsRLCK185 [28]. PBL27 also plays an important role for CERK1-mediated chitin signaling, though it seems not to contribute to FLS2-mediated flg22 signaling. Zhang et al. reported that BIK1, which is another cytoplasmic receptor-like kinase required for flagellin signaling, interacts with CERK1 and regulates chitin signaling in *Arabidopsis* [29].

For invading fungal pathogens, the presence of chitin-triggered immunity in plants is really a barrier to overcome. Recently, it has been shown that some pathogenic fungi evolved effectors to escape from the detection by chitin-mediated plant defense system to facilitate their invasion into the host plants. Fungal effectors, Ecp6 and Avr4, are known to inhibit chitin-triggered host immunity in different ways [30, 31]. Avr4 of *Cladosporium fulvum* is a chitin-binding protein and secreted into apoplast during infection. Avr4 then binds and protects chitin in the fungal cell wall from the degradation by host chitinase. Ecp6 produced by the same fungus is also a secreted protein and contains LysM motifs as similar to the plant chitin receptors. Ecp6 strongly binds chitin oligosaccharides released from the fungal cell walls during the infection process, resulting in the sequestration of chitin oligosaccharides from the detection by host plants. Ecp6-like LysM effectors were also found in other fungi, such as Slp1 from *Magnaporthe oryzae* and Mg3LysM from *Mycosphaerella graminicola* [32, 33]. The lack of these effectors results in the significant decrease of fungal/oomycete virulence, indicating the effectiveness of chitin-mediated defense system in plant immunity.

Peptidoglycan Receptor

Peptidoglycan (PGN) is a major cell wall component of gram-negative and gram-positive bacteria. PGN has a lattice structure formed from polymeric carbohydrate chains and oligopeptide bridges. These polymeric carbohydrate chains are consisted of alternately β-1,4-linked *N*-acetylglucosamine and *N*-acetylmuramic acid (MurNAc) residues. PGNs are located on the bacterial cell surface and regarded as a typical bacterial MAMP [34]. Indeed, these PGNs are recognized and induce defense responses in plants, insects, and animals [35–37]. In insects and animals, several proteins are known to be involved in the perception of PGNs. They include the family of peptidoglycan recognition proteins (PGRPs), nucleotide-binding oligomerization domain-containing proteins (NODs), and Toll-like receptor 2 (TLR2) [38]. On the other hand, whether the homologs of these proteins are involved in the perception of PGNs in plants is not known.

The first suggestion about plant PGN receptors came from the finding that the *cerk1* mutant showed an increased susceptibility to bacterial infection [39]. This result suggested that CERK1 is also involved in the perception of bacterial MAMPs. As the glycan chains of PGN and chitin are structurally very similar, consisting of GlcNAc/MurNAc or GlcNAc, respectively, and also the LysM motif was first identified in a PGN-degrading enzyme [40], it seemed natural to expect that PGN is a candidate of bacterial MAMP recognized by CERK1, though the possibility was questioned at the beginning [41]. Willmann et al. showed that two *Arabidopsis* CEBiP homologs, LYM1, LYM3, and CERK1 are involved in PGN perception in *Arabidopsis* [42]. Their mutant plants showed the decrease of sensitivity or insensitivity to PGN and also a high susceptibility to bacterial infection. Recombinant LYM1 and LYM3 proteins directly bound PGNs but CERK1 did not. These results and also the prediction that LYM1 and LYM3 do not have intracellular domain suggested that the *Arabidopsis* PGN perception system is analogous to the rice chitin receptor, consisting of corresponding binding protein(s) and a LysM receptor-like kinase that serves for membrane signaling. In the case of rice, homologs of LYM1 and LYM3, OsLYP4 and OsLYP6, were shown to be involved in the perception of PGN [23]. Considering the similarity of the glycan structure of PGN and chitin, and also the similarity of the LysM proteins involved in the perception of these MAMPs, it might be possible that their epitopes reside in the glycan backbone in both cases.

β-Glucan Receptor

β-Glucan oligosaccharides derived from fungal and oomycete cell walls are a major MAMP for this group of microbes [1]. These oligosaccharides seem to be generated from the fungal cell walls at the site of infection through the action of plant β-glucanases [43]. In addition, invading microbes themselves may generate such cell wall fragments, as observed in the spore germination of *Phytophthora sojae* [44]. While β-glucan poly- and oligosaccharides have been reported to show elicitor activity on a broad range of plant species, structures recognized by each plant seem to be different as discussed below.

A doubly branched hepta-β-glucoside is an example of a well-characterized β-glucan oligosaccharide elicitor, generated from *P. sojae* glucan [45]. From detailed studies with a set of structurally related oligosaccharides, it was shown that soybean cells recognize specific features of the hepta-β-glucoside structure, including all three non-reducing terminal glucosyl residues and their spacing along the backbone of the molecule [46]. For rice, a pentasaccharide purified from an enzymatic digest of β-glucan from the rice blast fungus, *Magnaporthe grisea* (*Pyricularia oryzae*), showed potent elicitor activity to induce phytoalexin biosynthesis [47]. Interestingly, these two oligosaccharides had a quite contrasting structure. The hepta-β-glucoside elicitor from *P. sojae* had the backbone of a 1,6-linked β-glucooligosaccharide with branches at the 3-position, while the elicitor-active β-glucopentaose from *M. grisea* had the backbone of 1,3-linked β-glucooligosaccharide branched at the 6-position. Comparison of the elicitor activity of the pentasaccharide from *M. grisea* and synthetic hexa-β-glucoside indicated that each glucan fragment can only be recognized by the corresponding host plant [47]. These results indicate clear differences in the specificity of the corresponding receptors in rice and soybean.

A binding protein for the hepta-β-glucoside elicitor was purified from soybean and named GBP [48, 49]. As GBP seems to have no intracellular domains, it has been suggested that it requires the presence of additional component(s) for signaling, as in the case of CEBiP in rice. In support of these considerations, several groups reported that GBP behaved as a high-molecular-weight protein complex, which might contain partner proteins of GBP [50, 51]. GBP was also shown to have endo β-1,3-glucanase activity, indicating that GBP has the ability not only to sense but also to generate β-glucan fragments [52].

It has been becoming clear that some pathogens change their cell wall components to evade the detection by plant hosts through the perception of cell wall-derived MAMPs. Rice blast fungus, *M. grisea*, starts to synthesize a novel α-1,3-glucan after the fungus detected wax components on the surface of host plant. The α-1,3-glucan seems to mask other cell wall components such as chitin and β-1,3-glucan and prevent their degradation by plant hydrolases [53]. Interestingly, similar strategy has been observed for animal pathogens. It was reported that a pathogenic yeast, *Histoplasma capsulatum*, synthesize α-1,3-glucan to prevent the detection of cell wall β-1,3-glucan by host receptor, dectin-1 [54]. Conversion of chitin to chitosan seems to be another strategy for pathogenic fungi to prevent their degradation by host chitinases and detection by host chitin receptors. Several phytopathogenic fungi were shown to expose chitosan on the cell surface of infection structures after invasion, while chitin became not detectable [55].

Oligogalacturonide Receptor

In addition to MAMPs derived from microbes, molecules generated from the host plants during pathogen infection also contribute for the detection of invading pathogens. These molecules are generally called as damage/danger-associated molecular patterns (DAMPs) or host-associated molecular patterns (HAMPs) [56, 57]. DAMPs are recognized by corresponding pattern recognition receptors (PRRs) and induce various immune responses as similar to MAPMs. One of the typical DAMP molecules is α-1,4-linked oligogalacturonides (OGs) that are derived from pectic polysaccharides of plant cell walls [2, 58–60]. Degradation of pectic polysaccharides by polygalacturonase (PGase) or pectin lyase secreted by pathogenic microbes generates these OGs. OGs released from the cell wall have been reported to induce various immune responses and enhance resistance to pathogens. On the other hand, many plants are known to produce polygalacturonase-inhibiting proteins (PGIPs) which contain leucine-rich repeat (LRR) domain [58, 61]. PGIPs are localized in the plant cell wall and have been proposed to attenuate the degradation of polygalacturonan backbone of pectic polysaccharides by pathogen-derived PGases, thus helping the accumulation of elicitor-active OGs. The importance of PGIPs in plant immunity has been shown by the analysis of defense responses as well as disease resistance of PGIP-overexpressing and knockout plants [62–64]. It seems possible that OGs and PGIPs enhance immune responses as proposed but PGIPs may also directly contribute to disease resistance through the inhibition of pathogenic PGases, which seriously damage host plants by their macerating activity.

Wall-associated kinase (WAK) family proteins have been thought to be the candidate of OG receptor. WAK family proteins are a plasma membrane protein consisting of an extracellular domain with EGF-like repeats, a transmembrane domain, and an intracellular Ser/Thr kinase domain. *Arabidopsis* WAK family proteins, WAK1 and WAK2, were shown to bind OGs and pectin in vitro [65, 66]. OGs were shown to bind to the non-EGF-like region in extracellular domain of WAK1 and five basic amino acid residues were shown to be involved in the OG binding [67, 68]. Finally, Brutus et al. showed that a chimeric receptor consisting of the extracellular domain of WAK1 and intracellular domain of EFR responded to

OGs and induced defense responses in the transfected tobacco leaves. Conversely, another chimeric receptor consisting of the extracellular domain of EFR and the intracellular domain of WAK1 responded to elf18 and induced defense responses in the transfected tobacco. These results showed the functionality of both extracellular and intracellular domains of WAK1 as a defense receptor, supporting the notion that WAK1 is the OG receptor [69].

References

1. Shibuya N, Minami E. Oligosaccharide signalling for defense responses in plant. Physiol Mol Plant Pathol. 2001;59:223–33.
2. Silipo A, Erbs G, Shinya T, Dow JM, Parrilli M, Lanzetta R, Shibuya N, Newman MA, Molinaro A. Glyco-conjugates as elicitors or suppressors of plant innate immunity. Glycobiology. 2010;20:406–19.
3. Barber MS, Bertram RE, Ride JP. Chitin oligosaccharides elicit lignification in wounded wheat leaves. Physiol Mol Plant Pathol. 1989;34:3–12.
4. Felix G, Regenass M, Boller T. Specific perception of subnanomolar concentrations of chitin fragments by tomato cells—induction of extracellular alkalinization, changes in protein-phosphorylation, and establishment of a refractory state. Plant J. 1993;4:307–16.
5. Yamada A, Shibuya N, Kodama O, Akatsuka T. Induction of phytoalexin formation in suspension-cultured rice cells by N-acetylchitooligosaccharides. Biosci Biotechnol Biochem. 1993;57:405–9.
6. Shibuya N, Kaku H, Kuchitsu K, Maliarik MJ. Identification of a novel high-affinity binding site for N-acetylchitooligosaccharide elicitor in the membrane fraction from suspension-cultured rice cells. FEBS Lett. 1993;329:75–8.
7. Baureithel K, Felix G, Boller T. Specific, high affinity binding of chitin fragments to tomato cells and membranes. Competitive inhibition of binding by derivatives of chitooligosaccharides and a Nod factor of Rhizobium. J Biol Chem. 1994;269:17931–8.
8. Shibuya N, Ebisu N, Kamada Y, Kaku H, Cohn J, Ito Y. Localization and binding characteristics of a high-affinity binding site for N-acetylchitooligosaccharide elicitor in the plasma membrane from suspension-cultured rice cells suggest a role as a receptor for the elicitor signal at the cell surface. Plant Cell Physiol. 1996;37:894–8.
9. Ito Y, Kaku H, Shibuya N. Identification of a high-affinity binding protein for N-acetylchitooligosaccharide elicitor in the plasma membrane of suspension-cultured rice cells by affinity labeling. Plant J. 1997;12:347–56.
10. Day RB, Okada M, Ito Y, Tsukada K, Zaghouani H, Shibuya N, Stacey G. Binding site for chitin oligosaccharides in the soybean plasma membrane. Plant Physiol. 2001;126:1162–73.
11. Okada M, Matsumura M, Ito Y, Shibuya N. High-affinity binding proteins for N-acetylchitooligosaccharide elicitor in the plasma membranes from wheat, barley and carrot cells: conserved presence and correlation with the responsiveness to the elicitor. Plant Cell Physiol. 2002;43:505–12.
12. Kaku H, Nishizawa Y, Ishii-Minami N, Akimoto-Tomiyama C, Dohmae N, Takio K, Minami E, Shibuya N. Plant cells recognize chitin fragments for defense signaling through a plasma membrane receptor. Proc Natl Acad Sci U S A. 2006;103:11086–91.
13. Shinya T, Motoyama N, Ikeda A, Wada M, Kamiya K, Hayafune M, Kaku H, Shibuya N. Functional characterization of CEBiP and CERK1 homologs in arabidopsis and rice reveals the presence of different chitin receptor systems in plants. Plant Cell Physiol. 2012;53:1696–706.
14. Fliegmann J, Uhlenbroich S, Shinya T, Martinez Y, Lefebvre B, Shibuya N, Bono JJ. Biochemical and phylogenetic analysis of CEBiP-like LysM domain-containing extracellular proteins in higher plants. Plant Physiol Biochem. 2011;49:709–20.

15. Tanaka S, Ichikawa A, Yamada K, Tsuji G, Nishiuchi T, Mori M, Koga H, Nishizawa Y, O'Connell R, Kubo Y. *HvCEBiP*, a gene homologous to rice chitin receptor *CEBiP*, contributes to basal resistance of barley to *Magnaporthe oryzae*. BMC Plant Biol. 2010;10:288.
16. Miya A, Albert P, Shinya T, Desaki Y, Ichimura K, Shirasu K, Narusaka Y, Kawakami N, Kaku H, Shibuya N. CERK1, a LysM receptor kinase, is essential for chitin elicitor signaling in *Arabidopsis*. Proc Natl Acad Sci U S A. 2007;104:19613–8.
17. Wan J, Zhang XC, Neece D, Ramonell KM, Clough S, Kim SY, Stacey MG, Stacey G. A LysM receptor-like kinase plays a critical role in chitin signaling and fungal resistance in *Arabidopsis*. Plant Cell. 2008;20(2):471–81.
18. Shimizu T, Nakano T, Takamizawa D, Desaki Y, Ishii-Minami N, Nishizawa Y, Minami E, Okada K, Yamane H, Kaku H, Shibuya N. Two LysM receptor molecules, CEBiP and OsCERK1, cooperatively regulate chitin elicitor signaling in rice. Plant J. 2010;64:204–14.
19. Wan J, Tanaka K, Zhang XC, Son GH, Brechenmacher L, Nguyen TH, Stacey G. LYK4, a lysin motif receptor-like kinase, is important for chitin signaling and plant innate immunity in Arabidopsis. Plant Physiol. 2012;160:396–406.
20. Liu T, Liu Z, Song C, Hu Y, Han Z, She J, Fan F, Wang J, Jin C, Chang J, Zhou JM, Chai J. Chitin-induced dimerization activates a plant immune receptor. Science. 2012;336:1160–4.
21. Faulkner C, Petutschnig E, Benitez-Alfonso Y, Beck M, Robatzek S, Lipka V, Maule AJ. LYM2-dependent chitin perception limits molecular flux via plasmodesmata. Proc Natl Acad Sci U S A. 2013;110:9166–70.
22. Narusaka Y, Shinya T, Narusaka M, Motoyama N, Shimada H, Murakami K, Shibuya N. Presence of LYM2 dependent but CERK1 independent disease resistance in *Arabidopsis*. Plant Signal Behav. 2013;8(9).
23. Liu B, Li JF, Ao Y, Qu J, Li Z, Su J, Zhang Y, Liu J, Feng D, Qi K, He Y, Wang J, Wang HB. Lysin motif-containing proteins LYP4 and LYP6 play dual roles in peptidoglycan and chitin perception in rice innate immunity. Plant Cell. 2012;24:3406–19.
24. Hayafune M, Berisio R, Marchetti R, Silipo A, Kayama M, Desaki Y, Arima S, Squeglia F, Ruggiero A, Tokuyasu K, Molinaro A, Kaku H, Shibuya N. Chitin-induced activation of immune signaling by the rice receptor CEBiP relies on a unique sandwich-type dimerization. Proc Natl Acad Sci U S A. 2014;111:E404–13.
25. Petutschnig EK, Jones AM, Serazetdinova L, Lipka U, Lipka V. The lysin motif receptor-like kinase (LysM-RLK) CERK1 is a major chitin-binding protein in *Arabidopsis thaliana* and subject to chitin-induced phosphorylation. J Biol Chem. 2010;285:28902–11.
26. Akamatsu A, Wong HL, Fujiwara M, Okuda J, Nishide K, Uno K, Imai K, Umemura K, Kawasaki T, Kawano Y, Shimamoto K. An OsCEBiP/OsCERK1-OsRacGEF1-OsRac1 module is an essential early component of chitin-induced rice immunity. Cell Host Microbe. 2013;13:465–76.
27. Yamaguchi K, Yamada K, Ishikawa K, Yoshimura S, Hayashi N, Uchihashi K, Ishihama N, Kishi-Kaboshi M, Takahashi A, Tsuge S, Ochiai H, Tada Y, Shimamoto K, Yoshioka H, Kawasaki T. A receptor-like cytoplasmic kinase targeted by a plant pathogen effector is directly phosphorylated by the chitin receptor and mediates rice immunity. Cell Host Microbe. 2013;13:347–57.
28. Shinya T, Yamaguchi K, Desaki Y, Yamada K, Narisawa T, Kobayashi Y, Maeda K, Suzuki M, Tanimoto T, Takeda J, Nakashima M, Funama R, Narusaka M, Narusaka Y, Kaku H, Kawasaki T, Shibuya N. Selective regulation of the chitin-induced defense response by the Arabidopsis receptor-like cytoplasmic kinase PBL27. Plant J. 2014;79:56–66.
29. Zhang J, Li W, Xiang T, Liu Z, Laluk K, Ding X, Zou Y, Gao M, Zhang X, Chen S, Mengiste T, Zhang Y, Zhou JM. Receptor-like cytoplasmic kinases integrate signaling from multiple plant immune receptors and are targeted by a *Pseudomonas syringae* effector. Cell Host Microbe. 2010;7:290–301.
30. van den Burg HA, Harrison SJ, Joosten MH, Vervoort J, de Wit PJ. *Cladosporium fulvum* Avr4 protects fungal cell walls against hydrolysis by plant chitinases accumulating during infection. Mol Plant Microbe Interact. 2006;19:1420–30.
31. de Jonge R, van Esse HP, Kombrink A, Shinya T, Desaki Y, Bours R, van der Krol S, Shibuya N, Joosten MHAJ, Thomma BPHJ. Conserved fungal LysM effector Ecp6 prevents chitin-triggered immunity in plants. Science. 2010;329:953–5.

32. Mentlak TA, Kombrink A, Shinya T, Ryder LS, Otomo I, Saitoh H, Terauchi R, Nishizawa Y, Shibuya N, Thomma BP, Talbot NJ. Effector-mediated suppression of chitin-triggered immunity by *Magnaporthe oryzae* is necessary for rice blast disease. Plant Cell. 2012;24:322–35.

33. Marshall R, Kombrink A, Motteram J, Loza-Reyes E, Lucas J, Hammond-Kosack KE, Thomma BP, Rudd JJ. Analysis of two in planta expressed LysM effector homologs from the fungus *Mycosphaerella graminicola* reveals novel functional properties and varying contributions to virulence on wheat. Plant Physiol. 2011;156:756–69.

34. Erbs G, Newman MA. The role of lipopolysaccharide and peptidoglycan, two glycosylated bacterial microbe-associated molecular patterns (MAMPs), in plant innate immunity. Mol Plant Pathol. 2012;13:95–104.

35. Felix G, Boller T. Molecular sensing of bacteria in plants. The highly conserved RNA-binding motif RNP-1 of bacterial cold shock proteins is recognized as an elicitor signal in tobacco. J Biol Chem. 2003;278:6201–8.

36. Gust AA, Biswas R, Lenz HD, Rauhut T, Ranf S, Kemmerling B, Gotz F, Glawischnig E, Lee J, Felix G, Nurnberger T. Bacteria-derived peptidoglycans constitute pathogen-associated molecular patterns triggering innate immunity in *Arabidopsis*. J Biol Chem. 2007;282:32338–48.

37. Boudreau MA, Fisher JF, Mobashery S. Messenger functions of the bacterial cell wall-derived muropeptides. Biochemistry. 2012;51:2974–90.

38. Sorbara MT, Philpott DJ. Peptidoglycan: a critical activator of the mammalian immune system during infection and homeostasis. Immunol Rev. 2011;243:40–60.

39. Gimenez-Ibanez S, Hann DR, Ntoukakls V, Petutschnig E, Lipka V, Rathjen JP. AvrPtoB targets the LysM receptor kinase CERK1 to promote bacterial virulence on plants. Curr Biol. 2009;19:423–9.

40. Butler AR, O'Donnell RW, Martin VJ, Gooday GW, Stark MJ. *Kluyveromyces lactis* toxin has an essential chitinase activity. Eur J Biochem/FEBS. 1991;199:483–8.

41. Gimenez-Ibanez S, Ntoukakis V, Rathjen JP. The LysM receptor kinase CERK1 mediates bacterial perception in Arabidopsis. Plant Signal Behav. 2009;4:539–41.

42. Willmann R, Lajunen HM, Erbs G, Newman MA, Kolb D, Tsuda K, Katagiri F, Fliegmann J, Bono JJ, Cullimore JV, Jehle AK, Gotz F, Kulik A, Molinaro A, Lipka V, Gust AA, Nurnberger T. *Arabidopsis* lysin-motif proteins LYM1 LYM3 CERK1 mediate bacterial peptidoglycan sensing and immunity to bacterial infection. Proc Natl Acad Sci U S A. 2011;108:19824–9.

43. Okinaka Y, Mimori K, Takeo K, Kitamura S, Takeuchi Y, Yamaoka N, Yoshikawa M. A structural model for the mechanisms of elicitor release from fungal cell walls by plant β-1,3-endoglucanase. Plant Physiol. 1995;109:839–45.

44. Waldmuller T, Cosio EG, Grisebach H, Ebel J. Release of highly elicitor-active glucans by germinating zoospores of *Phytophthora megasperma* f sp *glycinea*. Planta. 1992;188:498–505.

45. Sharp JK, McNeil M, Albersheim P. The primary structures of one elicitor-active and seven elicitor-inactive hexa(β-D-glucopyranosyl)-D-glucitols isolated from the mycelial walls of *Phytophthora megasperma* f. sp. *glycinea*. J Biol Chem. 1984;259:11321–36.

46. Cheong JJ, Birberg W, Fugedi P, Pilotti A, Garegg PJ, Hong N, Ogawa T, Hahn MG. Structure-activity relationships of oligo-β-glucoside elicitors of phytoalexin accumulation in soybean. Plant Cell. 1991;3:127–36.

47. Yamaguchi T, Yamada A, Hong N, Ogawa T, Ishii T, Shibuya N. Differences in the recognition of glucan elicitor signals between rice and soybean: β-glucan fragments from the rice blast disease fungus *Pyricularia oryzae* that elicit phytoalexin biosynthesis in suspension-cultured rice cells. Plant Cell. 2000;12:817–26.

48. Umemoto N, Kakitani M, Iwamatsu A, Yoshikawa M, Yamaoka N, Ishida I. The structure and function of a soybean β-glucan-elicitor-binding protein. Proc Natl Acad Sci U S A. 1997;94:1029–34.

49. Mithofer A, Fliegmann J, Neuhaus-Url G, Schwarz H, Ebel J. The hepta-β-glucoside elicitor-binding proteins from legumes represent a putative receptor family. Biol Chem. 2000;381:705–13.

50. Cheong JJ, Alba R, Cote F, Enkerli J, Hahn MG. Solubilization of functional plasma membrane-localized hepta-β-glucoside elicitor-binding proteins from soybean. Plant Physiol. 1993;103:1173–82.

51. Mithofer A, Lottspeich F, Ebel J. One-step purification of the β-glucan elicitor-binding protein from soybean (*Glycine max* L.) roots and characterization of an anti-peptide antiserum. FEBS Lett. 1996;381:203–7.
52. Fliegmann J, Mithofer A, Wanner G, Ebel J. An ancient enzyme domain hidden in the putative β-glucan elicitor receptor of soybean may play an active part in the perception of pathogen-associated molecular patterns during broad host resistance. J Biol Chem. 2004;279:1132–40.
53. Fujikawa T, Kuga Y, Yano S, Yoshimi A, Tachiki T, Abe K, Nishimura M. Dynamics of cell wall components of *Magnaporthe grisea* during infectious structure development. Mol Microbiol. 2009;73:553–70.
54. Rappleye CA, Eissenberg LG, Goldman WE. *Histoplasma capsulatum* α-(1,3)-glucan blocks innate immune recognition by the β-glucan receptor. Proc Natl Acad Sci U S A. 2007;104:1366–70.
55. El Gueddari NE, Rauchhaus U, Moerschbacher BM, Deising HB. Developmentally regulated conversion of surface-exposed chitin to chitosan in cell walls of plant pathogenic fungi. New Phytol. 2002;156:103–12.
56. Boller T, Felix G. A renaissance of elicitors: perception of microbe-associated molecular patterns and danger signals by pattern-recognition receptors. Annu Rev Plant Biol. 2009;60:379–406.
57. De Lorenzo G, Brutus A, Savatin DV, Sicilia F, Cervone F. Engineering plant resistance by constructing chimeric receptors that recognize damage-associated molecular patterns (DAMPs). FEBS Lett. 2011;585:1521–8.
58. De Lorenzo G, Ferrari S. Polygalacturonase-inhibiting proteins in defense against phyto-pathogenic fungi. Curr Opin Plant Biol. 2002;5:295–9.
59. Ferrari S, Galletti R, Pontiggia D, Manfredini C, Lionetti V, Bellincampi D, Cervone F, De Lorenzo G. Transgenic expression of a fungal endo-polygalacturonase increases plant resistance to pathogens and reduces auxin sensitivity. Plant Physiol. 2008;146:669–81.
60. Galletti R, Denoux C, Gambetta S, Dewdney J, Ausubel FM, De Lorenzo G, Ferrari S. The AtrbohD-mediated oxidative burst elicited by oligogalacturonides in Arabidopsis is dispensable for the activation of defense responses effective against *Botrytis cinerea*. Plant Physiol. 2008;148:1695–706.
61. De Lorenzo G, D'Ovidio R, Cervone F. The role of polygalacturonase-inhibiting proteins (PGIPs) in defense against pathogenic fungi. Annu Rev Phytopathol. 2001;39:313–35.
62. Powell AL, van Kan J, ten Have A, Visser J, Greve LC, Bennett AB, Labavitch JM. Transgenic expression of pear PGIP in tomato limits fungal colonization. Mol Plant Microbe Interact. 2000;13:942–50.
63. Ferrari S, Galletti R, Vairo D, Cervone F, De Lorenzo G. Antisense expression of the *Arabidopsis thaliana AtPGIP1* gene reduces polygalacturonase-inhibiting protein accumulation and enhances susceptibility to *Botrytis cinerea*. Mol Plant Microbe Interact. 2006;19:931–6.
64. Janni M, Sella L, Favaron F, Blechl AE, De Lorenzo G, D'Ovidio R. The expression of a bean PGIP in transgenic wheat confers increased resistance to the fungal pathogen *Bipolaris soro-kiniana*. Mol Plant Microbe Interact. 2008;21:171–7.
65. Decreux A, Messiaen J. Wall-associated kinase WAK1 interacts with cell wall pectins in a calcium-induced conformation. Plant Cell Physiol. 2005;46:268–78.
66. Kohorn BD, Johansen S, Shishido A, Todorova T, Martinez R, Defeo E, Obregon P. Pectin activation of MAP kinase and gene expression is WAK2 dependent. Plant J. 2009;60:974–82.
67. Decreux A, Thomas A, Spies B, Brasseur R, Van Cutsem P, Messiaen J. In vitro characterization of the homogalacturonan-binding domain of the wall-associated kinase WAK1 using site-directed mutagenesis. Phytochemistry. 2006;67:1068–79.
68. Cabrera JC, Boland A, Messiaen J, Cambier P, Van Cutsem P. Egg box conformation of oligogalacturonides: the time-dependent stabilization of the elicitor-active conformation increases its biological activity. Glycobiology. 2008;18:473–82.
69. Brutus A, Sicilia F, Macone A, Cervone F, De Lorenzo G. A domain swap approach reveals a role of the plant wall-associated kinase 1 (WAK1) as a receptor of oligogalacturonides. Proc Natl Acad Sci U S A. 2010;107:9452–7.

Chapter 4
Marine Algae Oligo-carrageenans (OCs) Stimulate Growth and Defense Responses in Terrestrial Plants

Alejandra Moenne

Abstract Oligo-carrageenans (OCs) are obtained by acid hydrolysis of pure commercial carrageenans extracted from marine red macroalgae. OCs are constituted by around 20 units of sulfated galactose and, in some cases, anhydrogalactose units. The principal OCs are kappa, lambda, and iota, which have been observed to stimulate growth and defense responses in plants. Regarding growth stimulation efficiency, the best effect in tobacco plants was obtained with OCs iota and kappa, in *Eucalyptus globulus* trees with OCs kappa and iota, and in *Pinus radiata* trees with OC kappa. Regarding defense stimulation efficiency, the best effect was obtained in tobacco plants with OCs lambda and iota and in *Eucalyptus* trees with OC kappa. OCs increased the amount of essential oils (volatile terpenoids) with potential repellent and insecticidal activities in *Eucalyptus* as well as the level of polyphenolic compounds with potential antimicrobial activities in tobacco plants. Regarding the mechanisms involved in growth stimulation, OCs increase net photosynthesis, basal metabolism, and cell division in tobacco plants, and in *Eucalyptus* and pine trees they mediate the increase in photosynthesis, C, N, and S assimilation, basal metabolism, and the level of growth-promoting hormones. Moreover, OC kappa increases ascorbate, glutathione, and NADPH syntheses, inducing a more reducing redox status; the increase in NADPH enhances thioredoxin reductase (TRR) and thioredoxin (TRX) activities which, in turn, stimulate photosynthesis, and basal and secondary metabolism in *Eucalyptus* trees. Similar OC-induced mechanisms may be occurring in tobacco plants and pine trees determining a reducing redox status that stimulate growth of NADH-TRR/TRX system, and most likely in other plants of agronomical and forestry importance.

Keywords Ascorbate • Defense responses • *Eucalyptus* • Glutathione • Growth • NADPH • Oligo-carrageenans • *Pinus radiata* • Reducing redox status • Thioredoxin system • Tobacco plants

A. Moenne, Ph.D. (✉)
Department of Biology, Faculty of Chemistry and Biology, University of Santiago of Chile,
Av. Libertador Bernardo O'Higgins 3363, Santiago, Región Metropolitana 9170022, Chile
e-mail: alejandra.moenne@usach.cl

© Springer Science+Business Media New York 2016   41
H. Yin, Y. Du (eds.), *Research Progress in Oligosaccharins*,
DOI 10.1007/978-1-4939-3518-5_4

Introduction

It is now clearly established that plant and marine algae oligosaccharides induce stimulation or inhibition of growth and development in plants [1–7]. In addition, plant and marine algae oligosaccharides stimulate defense responses, increasing protection against pathogens in plants [2, 5, 8–11]. The best studied plant oligosaccharides and their effects in plants are oligo-galacturonides (OGs), which are derived from homogalacturonan, the major component of pectin in plant cell walls [12]. It was determined that OGs bind to a specific plasma membrane receptor corresponding to wall-associated kinase 1 (WAK1), which is constituted by an extracellular domain with several epidermal growth factor (EGF)-like repeats and an N-terminal non-EGF domain that binds OGs [13]. The binding domain is coupled to an intracellular Ser/Thr kinase domain that activates signal transduction triggering an oxidative burst; an increase in nitric oxide, jasmonic acid, salicylic acid, and ethylene; and the activation of MAPK signaling that activates defense gene expression [10, 14]. While OGs stimulate defense responses protecting plants against pathogens [15, 16] they inhibit growth since they have an antagonistic effect with auxins [17]. It has been shown that OGs inhibit activation of auxin-dependent genes at transcriptional level probably interacting with transcription factors ARF, since they do not induce the stabilization of auxin-response repressors or the decrease of auxin receptor level [18]; for a model, see Ferrari et al. [12].

In regard to marine algae oligosaccharides, no specific plasma membrane receptor has been cloned so far; indeed, the mechanisms involved in the stimulation of growth and defense in plants are just being further elucidated. The best studied marine algae oligosaccharides are oligo-carrageenans (OCs) and, in contrast to OGs, they have been observed to simultaneously stimulate both growth and defenses in plants (see next sections).

Discovery of OC-Induced Stimulation of Growth

While studying the stimulation of defense responses in tobacco plants (var. Xanthi) induced by OCs, it appears that plants treated with OCs kappa1, kappa2, lambda, and iota at concentrations of 1 mg/mL, once a week, four times in total, and cultivated without additional treatment for 45 days, showed an increase in height and in number of leaves compared to control plants (Fig. 4.1), mainly with OCs kappa1 and kappa2 [19]. The increase in height in plants treated with OCs kappa and iota was 3.7- and 3-fold, respectively, and the increase in foliar biomass was about 2-fold. The stimulation of growth was obtained using OC concentrations from 0.5 to 5 mg/mL, since higher concentrations induced an inhibitory effect of growth. In addition, OCs induced protection against tobacco mosaic virus (TMV) in tobacco plants (var. Xanthi), mainly OC lambda that decreased the number of necrotic lesions in almost 95 % [19].

Fig. 4.1 Tobacco plants (var. Xanthi) sprayed on leaves with water (control, C) or with an aqueous solution of OCs kappa1 (K1), kappa2 (K2), lambda (L), and iota (I), at a concentration of 1 mg/mL, once a week, four times in total, and cultivated without additional treatment for 45 days

Preparation of OCs

Pure (free of secondary metabolites) commercial carrageenans kappa2, lambda, and iota (20 g) were purchased from Gelymar S.A. (Santiago, Chile) and were solubilized in 2 L of water at 60 °C. Concentrated HCl (36.2 N) was added to reach a final concentration of 0.1 N, the solution was incubated for 45 min at 60°, and NaOH 1 M was added to obtain a neutral pH. A sample of 10 µL of each depolymerized carrageenan (oligo-carrageenan) was analyzed by electrophoresis in an agarose gel (1.5 % w/v) using 100 V for 1 h and dextran sulfate of 8 and 10 kDa as standarts (Sigma, St. Louis, USA). The gel was stained with 15 % w/v alcian blue dye in 30 % v/v acetic acid for 1 h at room temperature and washed with 50 % v/v acetic acid for 1 h. Oligo-carrageenans kappa2, lambda, and iota were visualized as a relative discrete band of around 10 kDa (Fig. 4.2).

Structure of OCs

Oligo-carrageenans kappa2, lambda, and iota are constituted by around 20 units of sulfated galactose linked by alternate β-1,4 and α-1,3 glycosidic bonds with sulfate groups located in positions 2, 4, and 6 of the galactose ring and with

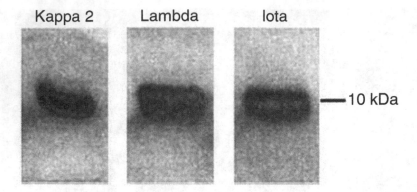

Fig. 4.2 Oligo-carrageenans (OCs) kappa2, lambda, and iota are visualized as a relative discrete band of around 10 kDa (taken from Vera et al. [27])

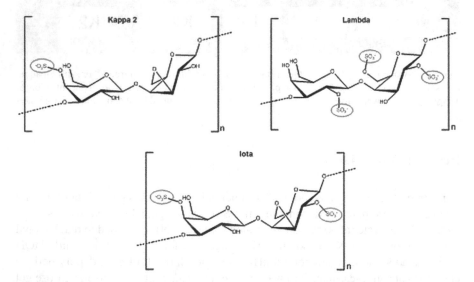

Fig. 4.3 Structure of the disaccharide ($n = 5$) that constitutes OCs kappa2, lambda, and iota which contain sulfated galactose units, and anhydrogalactose units in the case of OCs kappa2 and iota (taken from Vera et al. [11])

anhydrogalactose units in some cases (Fig. 4.3). In particular, oligo-carrageenan kappa has one sulfate group per disaccharide unit and an anhydrogalactose residue, oligo-carrageenan lambda has three sulfate groups per disaccharide unit, and oligo-carrageenan iota presents two sulfate groups per disaccharide unit and an anhydrogalactose residue. Thus, the degree of sulfation increases in oligo-carrageenans kappa, iota, and lambda and only oligo-carrageenans kappa and iota have anhydrogalactose residues.

Treatment with OCs

Plants ($n = 10$ for each group) with an initial height of 25–30 cm were sprayed in the upper and lower part of the leaves with 2–5 mL of water per plant (control group) or with 2–5 mL of an aqueous solution of OCs kappa, lambda, or iota (treated groups 1, 2, and 3) at a concentration of 1 mg/mL, once a week, four times in total, and plants were cultivated for months without additional treatment, outdoors in plastic bags with compost, or in the field.

OC-Induced Stimulation of Plant Growth and Involved Mechanisms

The effect of OCs in the stimulation of growth was further studied in commercial tobacco plants (var. Burley). The best growth stimulatory effect was obtained with OC iota, which induced a 2.5 times increase in leaf biomass [20]. It was shown that net photosynthesis increased in response to all three OCs, but the activity of ribulose 1,5 biphosphate carboxylase/oxygenase (rubisco), involved in C assimilation, was higher in plants treated with OC iota (around four times), as well as the activity of glutamate dehydrogenase, involved in N assimilation (around threefold). In addition, the level of transcripts of cyclins A and D and cyclin-dependent kinases, CDKA and CDKB, involved in cell cycle regulation, was higher in OC iota-treated plants in around 3.5 to 5 times. Thus, higher C and N assimilation and cell division may explain increased growth stimulation induced by OC iota in tobacco plants (var. Burley).

The effects of OCs on growth stimulation were also analyzed in *Eucalyptus globulus* trees cultivated for 1 year in plastic bags with compost, and for 3 years in the field [21]. It was shown that the best effect was obtained with OCs kappa and iota, which induced a 58 % and 47 % increase in height, respectively, and a 44 % and 40 % increase in trunk diameter, respectively [21]. Net photosynthesis increased in *Eucalyptus* trees treated with OCs kappa and iota in 23 % as well as rubisco activity, involved in C assimilation; glutamine synthase (GlnS) activity, involved in N assimilation; and adenosine 5′ phosphosulfate reductase (APR) and O-acetylserine thiol-lyase (O-ASTL) activities, involved in S assimilation [22]. Furthermore, the content of α-cellulose increased in *Eucalyptus* treated with OCs kappa and iota in 16 % and 13 %, respectively [21]. Thus, OCs induced a double-beneficial effect by increasing both wood volume and α-cellulose content in *Eucalyptus* trees.

In addition, the effect of OC kappa was analyzed in *Pinus radiata* trees cultivated in plastic bags with compost for 9 months [23]. OC kappa was assayed at concentrations of 1 and 5 mg/mL, and both concentrations were observed to enhance rubisco, GlnS, and APR activities, involved in C, N, and S assimilation, respectively, as well as in the level of the growth-promoting hormones auxin indole 3-acetic acid (IAA) and gibberellin A_3 (GA_3), although the latter mainly when OC kappa was used at a concentration of 1 mg/mL. Pine trees were transferred to the field to be monitored for the next three more years in order to determine a potential increase of α-cellulose content, as observed in *Eucalyptus* trees.

In order to analyze the mechanism determining the stimulation of growth, *Eucalyptus* trees were treated with OC kappa and cultivated in plastic bags with compost for 4 months and the level of reducing compounds such as ascorbate (ASC), glutathione (GSH), and NADPH was determined as well as the activities of thioredoxin reductase (TRR) and thioredoxins (TRXs) [22]. TRR/TRX system is involved in the regulation of several metabolic processes such as photosynthesis, and basal and secondary metabolism [24, 25]. In addition, trees were treated with inhibitors of ASC, GSH, and NAD(P)H syntheses as well as of TRR activity and with OC kappa, and photosynthesis, basal metabolism, and growth were analyzed. It was shown that OC kappa induced an increase in ASC, GSH, and NADPH syntheses as well as in TRR/TRX activities, indicating that the redox status was changed to a more reducing condition in *Eucalyptus* trees. Treatment with inhibitors showed that the increase in ASC, GSH, and NADPH is a cross-talking factor and that the increase in NADPH activates TRR/TRX system, and that TRR/TRX system activates photosynthesis, C, N, and S assimilation, basal metabolism, and growth in *Eucalyptus* trees which may explain, at least in part, the increase in growth [22].

In addition, the level of growth-promoting hormones, IAA, GA3, and the cytokinin *trans*-zeatin (*t*-Z), was also increased treated with OC kappa which may also determine the stimulation of growth and development in *Eucalyptus* trees [26]. Furthermore, the effect of inhibitors indicates that the increase in reducing compounds and in TRR/TRX activities participates in the increase of the growth-promoting hormone levels, mainly IAA, GA_3, and *t*-Z [26].

OC-Induced Stimulation of Defense Responses and Involved Mechanisms

OCs kappa, iota, and lambda at a concentration of 1 mg/mL induced protection against tobacco mosaic virus (TMV) in tobacco plants (var. Xanthi) [19, 27]. In addition, OCs induced protection against the infection by the fungus *Botrytis cinerea*, mainly by OCs lambda and iota, and by the bacterium *Pectobacterium carotovorum* [27]. Moreover, OCs induced a partial suppression of viral infection as well as a complete suppression of fungal or bacterial infections [27]. It was shown that OCs activate the phenylpropanoid pathway in tobacco plants leading to an increase in the level of several phenylpropanoid compounds (PPCs), among which salicylic acid, dihydrobenzoic acid, chlorogenic acid, ferulic acid, scopoletin, esculetin, kaempferol, and quercetin, having antimicrobial activities in vitro, were identified [27]. Thus, the increase in protection against microbial infections and the suppression of these infections are probably due to the increase in PPCs having antimicrobial activities.

On the other hand, OCs kappa, iota, and lambda induced an increase in the level of essential oils (volatile terpenoids) in *Eucalyptus* trees cultivated in the field for 3 years [21], mainly OCs kappa and iota which increased total essential oils in 67 % and 39 %, respectively. In addition, the level of total essential oils increased in 22 % after cultivating *Eucalyptus* trees with OC kappa only for 4 months [28]. Treatment with OC kappa induced a decrease in the main terpenoids present in *Eucalyptus* essential oils, such as eucalyptol, sabinene, α-terpineol, α-pinene, δ-cadinene, and isoledene. In addition, OC kappa increased some minor terpenoids, such as silvestrene, α-pellandrene, γ-terpinene, β-pinene, γ-cadinene, aromadendrene, viridiflorene, α-gurjunene, γ-gurjunene, α-guaiene, and myrcene, which have been observed to have repellent and insecticidal activities in vitro [28]. Moreover, some newly synthesized terpenoids were observed in response to OC kappa, such as carene, α-terpinene, α-fenchene, γ-maaliene, and spathulenol, all of which have proven to provide repellent and insecticidal activities in vitro [28]. Thus, the increase of volatile terpenoids having repellent and insecticidal activities suggests that defense against insects should be increased in OC kappa-treated *Eucalyptus* trees; however, this assumption must be further tested experimentally.

Additional Features and Perspectives

It was also analyzed whether OC kappa can induce an increase in the levels of taxanes in *Taxus baccata* trees at concentration of 1 and 5 mg/mL [23]. Taxanes are complex diterpene molecules involved in defense against pathogens in *Taxus* [29, 30], and paclitaxel (Taxol) is one of the best anticancer drugs known to date to treat different types of human cancers, such as ovarian, mammary, and lung cancers [31]. *Taxus* trees were cultivated in plastic bags with compost for 1 year and then in the field for an additional year. Interestingly, the level of some taxanes corresponding to deacetyl-baccatine III, baccatine III, and placlitaxel (Taxol), the last compounds of taxane synthesis pathway, increased in *Taxus* leaves after 1 year but OC kappa did not increase height in *Taxus baccata* [23].

In the future, molecular mechanisms involved in the stimulation of growth and defense against pathogens will be analyzed in *Eucalyptus* trees. In particular, it is possible that a master gene controlling growth, and basal and secondary metabolisms, could be activated in response to the binding of OC kappa to its potential receptor. The latter gene could correspond to target of rapamycin (TOR), a protein kinase that controls sugar and basal metabolism in mammals and yeast [32], as well as secondary metabolism in plants [33]. Thus, the level of transcripts encoding TOR kinase as well as the level of the protein and its phosphorylation (active state) will be analyzed in control and OC kappa-treated *Eucalyptus* trees by our research group.

References

1. Albersheim P, Darvill A, Augur C, Cheong JJ, Eberhard S, Hahn MG, Marfá V, Mohnen D, O'Neill MA, Spiro MD, York WS. Oligosaccharins: oligosaccharide regulatory molecules. Acc Chem Res. 1992;25:77–83.
2. Darvill A, Bergmann C, Servone D, De Lorenzo G, Ham KS, Spiro MD, York WS, Albersheim P. Oligosaccharins involved in plant growth and host-pathogen interactions. Biochem Soc Symp. 1994;60:89–94.
3. Bellicampi D, Cardarelli M, Zaghi D, Serino G, Salvi G, Gatz C, Cervone F, Altamura MM, Constantino P, De Lorenzo G. Oligogalacturonides prevent rhizogenesis in rolB-transformed tobacco explants by inhibiting auxin-induced expression of rolB gene. Plant Cell. 1996;8:477–87.
4. Kollarová K, Henselová M, Lisková D. Effects of auxin and plant oligosaccharides on root formation and elongation growth of mung bean hypocotyls. J Plant Growth Regul. 2005;46:1–9.
5. Laporte D, Vera J, Chandía NP, Zúñiga EA, Matsuhiro B, Moenne A. Structurally unrelated algal oligosaccharides differentially stimulate growth and defense against tobacco mosaic virus in tobacco plants. J Appl Phycol. 2007;19:79–88.
6. Kollarová K, Zelko I, Henselová M, Kapek D, Lisková D. Growth and anatomical parameters of adventitious roots formed on mung bean hypocotyls are correlated with galactoglucomannan oligosaccharide structure. Sci World J. 2012. ID 797815, 7p.
7. González A, Castro J, Vera J, Moenne A. Seaweed oligosaccharides stimulate plant growth by enhancing carbon and nitrogen assimilation, basal metabolism, and cell division. J Plant Growth Regul. 2013;32:443–8.
8. Slovaková L, Lisková D, Capek P, Kuvocková Kakoniová D, Karacsonyi S. Defense responses against TNV infection induced by galactoglucomannans-derived oligosaccharides in cucumber cells. Eur J Plant Pathol. 2000;106:543–53.
9. Aziz A, Gauthier A, Bézier A, Poinssot B, Joubert JM, Pugin A, Heyraud A, Baillieul F. Elicitor and resistance-inducing activities of 1,4 cellodextrins in grapevine, comparison with β-1,3 glucan and α-1,4 oligogalacturonides. J Exp Bot. 2007;58:1463–72.
10. Denoux C, Galetti R, Mamarella L, Gopalan S, Werck D, De Lorenzo G. Activation of defense response pathway by OGs and Flg22 elicitors in *Arabidopsis* seedlings. Mol Plant. 2008;1:423–45.
11. Vera J, Castro J, González A, Moenne A. Seaweed polysaccharides and derived oligosaccharides stimulate defense responses and protection against pathogens in plants. Mar Drugs. 2011;9: 2514–25.
12. Ferrari S, Savatin DV, Sicilia F, Gramegna G, Servone F, De Lorenzo G. Oligo-galacturonides : plant damage-associated molecular patterns and regulators of growth and development. Front Plant Sci. 2013;4:1–9.
13. Brutus A, Sicilia F, Maccone A, Cervone F, De Lorenzo G. A domain swap approach reveals a role of the plant wall associated kinase 1 (WAK1) as a receptor of oligo-galacturonides. Proc Natl Acad Sci U S A. 2010;107:9452–7.
14. Galetti R, Ferrari S, De Lorenzo G. *Arabidopsis* MAPK3 and MPK6 Play different roles in basal, oligo-galacturonide- or flagellin-induced resistance against *Botrytis cinerea*. Plant Physiol. 2011;157:804–14.
15. Aziz AA, Heyraud A, Lambert B. Oligogalacturonide signal transduction, induction of defense-related responses and protection against *Botrytis cinerea*. Planta. 2004;218:767–74.
16. Ferrari S, Galetti R, Denoux C, De Lorenzo G, Ausubel FM, Dewdney J. Resistance to *Botrytis cinerea* induced in *Arabidopsis* by elicitors is independent of salicylic acid, ethylene or jasmonate signaling but requires PHYTOALEXIN DEFICIENT3. Plant Physiol. 2007;144:367–79.
17. Ferrari S, Galetti R, Pontiggia D, Manfredini C, Lionetti V, Bellicampi D, De Lorenzo G. Transgenic expression of a fungal *endo*-polygalacturonase increases plant resistance to pathogens and reduces auxin sensitivity. Plant Physiol. 2008;146(157):804–14.

18. Savatin DV, Ferrari S, Sicilia F, De Lorenzo G. Oligo-galacturonide-auxin antagonism does not require post-transcriptional genes silencing or stabilization of auxin response repressors in *Arabidopsis*. Plant Physiol. 2011;157:1163–74.
19. Moenne A. Composition and method to stimulate growth and defense against pathogens in plants. US Patent Application, number US2010/0173779 A1, 2010.
20. Castro J, Vera J, González A, Moenne A. Oligo-carrageenans stimulate growth by enhancing photosynthesis, basal metabolism, and cell cycle in tobacco plants (var. Burley). J Plant Growth Regul. 2012;31:173–85.
21. González A, Contreras RA, Moenne A. Oligo-carrageenans enhance growth and contents of cellulose, essential oils and polyphenolic compounds in *Eucalyptus globulus* trees. Molecules. 2013;18:8740–51.
22. González A, Moenne F, Contreras RA, Gómez M, Sáez CA, Moenne A. OC kappa increases NADPH, ascorbate and glutathione levels and TRR/TRX activities enhancing photosynthesis, basal metabolism, and growth in *Eucalyptus globulus*. Front Plant Sci. 2014;13:512. doi:10.3389/fpls.2014.00512.
23. Saucedo S, Contreras RA, Moenne A. OC kappa-induced increase in C, N and S assimilation, auxin and gibberellin contents, and growth in *Pinus radiata* trees. J Forest Res., in press
24. Gelhaye E, Rouhier N, Navrot N, Jacquot JP. The plant thioredoxin system. Cell Mol Life Sci. 2005;62:24–35.
25. Montrichard F, Alkhalfioui F, Yano H, Vensel WH, Hurkman WJ, Buchanan BB. Thioredoxin targets in plants: the first 30 years. J Proteomics. 2009;72:452–4.
26. González A, Contreras RA, Zúñiga G, Moenne A. Oligo-carrageenan kappa-induced reducing redox status and activation of TRR/TRX system increase the level of indole-3-acetic acid, gibberellin A₃ and *trans*-zeatin in *Eucalyptus globulus* trees. Molecules. 2014;19:12690–8.
27. Vera J, Castro J, González A, Moenne A. Oligo-carrageenans induce a long-term and broad range protection against pathogens in tobacco plants (var. Xhanti). Physiol Mol Plant Pathol. 2012;79:31–9.
28. González A, Gutiérrez-Cutiño M, Moenne A. Oligo-carrageenan kappa-induced reducing redox status and increase in TRR/TRX activities promote activation and reprogramming of terpenoid metabolism in *Eucalyptus* trees. Molecules. 2014;19:7356–67.
29. Croteau R, Ketchum REB, Long RN, Kaspera R, Wildung MR. Taxol biosynthesis and molecular genetics. Phytochem Rev. 2006;5:75–97.
30. Keeling CI, Bohlman J. Genes, enzymes and chemicals of terpenoid diversity in the constitutive and induced defense of conifers against insects and pathogens. New Phytol. 2006;170:657–75.
31. Fauzee NJS, Dong Z, Wang YL. Taxanes: promising anticancer drugs. Asian Pac J Cancer Prev. 2011;12:837–50.
32. Xiong Y, Sheen J. The role of target of rapamycin signaling network in plant growth and metabolism. Plant Physiol. 2014;164:499–512.
33. Xiong Y, McCormack M, Li L, Hall Q, Xiang C, Sheen J. Glucose-TOR signalling reprograms the transcriptome and activates meristems. Nature. 2013;496:181–6.

Chapter 5
Fructooligosaccharides: Effects, Mechanisms, and Applications

Moran Guo, Guochuang Chen, and Kaoshan Chen

Abstract Fructooligosaccharide (FOS) is a generic term for a series of homologous oligosaccharides in plants, composed of linear chains of fructose units, linked hy β (2 → 1) bonds. As one of the most widely commercially available prebiotics, the health benefits of dietary FOS have long been appreciated. Numerous experimental studies have demonstrated the roles of FOS in boosting immunity, reducing the risk and severity of gastrointestinal infection, inflammation, diarrhea, inflammatory bowel disease, obesity related metabolic disorders, and promoting anticancerous effect. However, little is known about their effect on inducing resistance in plants. In this chapter, we mainly introduce the induced resistance of burdock fructooligosaccharide (BFO), which is one of the most intensively studied FOS, in plants and postharvest fruits. As a potential elicitor, BFO could modulate the expression of defense-related genes and accumulation of secondary metabolites, especially salicylic acid-mediated pathway, related with multiple signaling pathways and defense components to enhance host defense responses in plants. A variety of applications in food formulations, medical treatment, and agriculture are also discussed.

Keyword Fructooligosaccharides • Elicitor • Burdock fructooligosaccharide • Induced resistance • Prebiotic

M. Guo, Ph.D.
Shijiazhuang Center for Disease Control and Prevention, Shijiazhuang 050000, China
e-mail: guowanpiaopeng0823@163.com

G. Chen, Ph.D.
School of Life Science and National Glycoengineering Research Center, Shandong University, Jinan 250100, China
e-mail: chenguochuang@126.com

K. Chen, Ph.D. (✉)
School of Life Science and National Glycoengineering Research Center, Shandong University, Jinan 250100, China

Department of Pharmacy, Wannan Medical College, Wuhu 241000, China

State Key Laboratory of Microbial Technology, Shandong University, Jinan 250100, China
e-mail: ksc313@126.com

© Springer Science+Business Media New York 2016
H. Yin, Y. Du (eds.), *Research Progress in Oligosaccharins*,
DOI 10.1007/978-1-4939-3518-5_5

Introduction

Oligosaccharides can be classified according to their chemical constituents and degree of polymerization, such as mannooligosaccharides, xylooligosaccharides, galactooligosaccharides, fructooligosaccharides, and so on [1–3]. Fructooligosaccharides (FOS), also sometimes called oligofructose or oligofructan, is a generic term for a series of homologous oligosaccharides with β $(2 \rightarrow 1)$ fructosyl-fructose glycosidic bonds. FOS have a number of interesting properties, including low sweetness intensity, non-cariogenicity, and are considered a source of soluble dietary fiber. Furthermore, FOS have important beneficial physiological effects such as low carcinogenicity, a prebiotic effect, improved mineral absorption, and decreased levels of serum cholesterol, triacylglycerols, and phospholipids. Currently FOS are increasingly included in food products and infant formulas due to their prebiotic effect—stimulating the growth of nonpathogenic intestinal microflora.

Source and Structure

FOS are considered as carbohydrates with a very low degree of polymerization and low molecular weight [4]. Structurally, FOS can be linear or branched fructose polymers and are generally denoted as Fn or GFn (G referring to the terminal glucose unit, F referring to fructose units, and n designating the number of fructose units in the fructan chain) [5]. As shown in Fig. 5.1, a FOS consists of a sucrose molecule to which other molecules of fructose have been added. The number of fructose residues in FOS is somewhat ambiguous. Many researchers agree with the view that FOS have a polymerized chain with n being 1 to 12 units of fructose. FOS also have been considered as polymers which include monosaccharide units between 2 and 20. However, the dividing point between oligo- and poly-fructooligosaccharides is 10 according to IUB-IUPAC terminology [6].

FOS are extracted from plants, fruits, and vegetables, such as bananas, onions, rye, leeks, wheat, Jerusalem artichoke, burdock, and yacón [7–9]. Jerusalem artichoke, chicory, and yacón have been found to have the highest concentrations of FOS. Yacón (*Smallanthus sonchifolius, Asteraceae*) is a tuber crop originally cultivated in South America in the Andean. Each yacón root typically stores carbohydrates (90–130 g/kg) [10], and the majority of the carbohydrates are inulin-type oligofructans and β-$(2 \rightarrow 1)$-fructooligosaccharides [11]. Some bacteria and fungi, such as *Penicillium expansum*, also have potential for sucrose conversion to FOS [12].

FOS can be obtained by extraction from plants (e.g., *Cichorium intybus* and *Arctium lappa*), based on water or aqueous alcohol, either methanol or ethanol extraction [7, 13]. The extraction was analyzed by high performance anion exchange chromatography-pulsed amperometric detection (HPAEC-PAD) and gas chromatography–mass spectrometry (GC-MS). FOS also can be partially hydrolyzed from purified inulin, using citric or phosphoric acids (pH, 2.0–2.5) as mild acid cata-

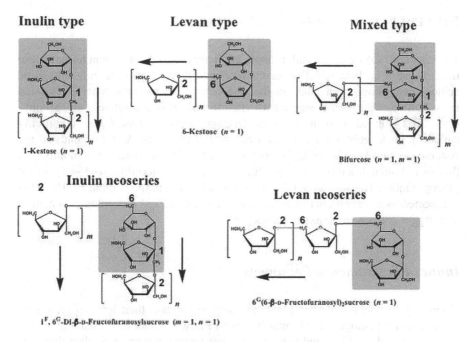

Fig. 5.1 Molecular structures of the different types of FOS (Benkeblia 2013) [6]

lysts [14]. In addition, FOS also could be enzymatically synthesized (transglycosylation of fungal enzyme beta-fructosidase) from sucrose [15, 16]. Now FOS are produced by continuous methods using immobilized cells, enzymes, membrane reactor systems, and microfiltration-based bioreactors with industrial feasibility [17].

Effect on Human and Animal Health

FOS are one of the most widely commercially available prebiotic compounds. FOS selectively stimulate the growth of specific microorganisms in the colon (e.g., *bifidobacteria*, *lactobacilli*) with a general positive health effect, and are known to decrease clinically relevant pathogenic germs [18]. They show a variety of pharmacological effects such as growth inhibition of microbial cells, reduction of cancer risk, scavenging of free radicals, and protection against cardiovascular disease. A double masked, randomized controlled study clearly demonstrated that FOS supplement decreased the occurrence of febrile illness and lipid homeostasis, activated anti-inflammatory activity, improved glucose homeostasis, and modulated certain aspects of the colonic and systemic immune system [19]. In addition, many studies have shown some benefits of FOS in the management of colorectal cancer [20]. However, any direct effect of FOS on the immune system cannot be excluded. Further studies on the efficacy of FOS in the prevention of infectious diseases should be performed.

Effect of FOS on Plants

The effect of FOS on health and its application are the focus of current research, but little research on FOS effect on plants has been done. Up to present, only some achievements have been made in the research of fructooligosaccharide isolated from burdock. Burdock fructooligosaccharide (BFO) is a plant reserve carbohydrate, first isolated from the roots of *Arcitum lappa. Arcitum lappa*, commonly called burdock, gobō (in Japan), is a biennial plant of the Arctium genus of the Asteraceae family. The BFO is composed of a linear chain of a number of β-(2 → 1)-linked fructofuranose residues with a single terminal α-(1 → 2)-linked glucopyranose. The amount of BFO in the air-dried root tissue is about 17.0%. BFO is reported as a natural elicitor which could induce marked physiological changes and trigger defense responses in plants [21].

Induced Resistance to Pathogens

Upon appropriate stimulation, plants are able to increase their level of resistance against future pathogen attack, which is well known as induced resistance. Some studies found that BFO could induce resistance against a number of plant diseases, although it has no antimicrobial activities in vitro [21, 22]. It could enhance resistance to *Colletotrichum lagenarium* and *Sphaerotheca fuliginea* in cucumber plants [22, 23]. The disease severity of *Lycopersicon esculentum* was evaluated 48 h after treatment with 0.6% BFO, followed by inoculation with a spore suspension of *Botrytis cinerea*. The results demonstrated that the disease index in BFO-treated plants was decreased by 42.5% compared with the control at 96 h [24]. In addition, BFO not only controls fungal diseases but also induces resistance in tobacco against tobacco mosaic virus [21].

Induced Chilling Resistance

BFO obviously promoted the growth and root development of cucumber seedlings and increased the cell membrane stability. An experiment at lower temperature suggests that BFO could promote the growth and increase the chilling resistance of cucumber seedlings to a certain extent [23].

Induced Synthesis of Secondary Metabolites

Secondary metabolites cover a very wide variety of unrelated micromolecule compounds, which regulate many physiological responses in plant growth, development and adaptation, and often play an important role in plant defense against pathogens [25]. Treatment with BFO evoked the activities of phenylalanine ammonia lyase and peroxidases to enhance the biosynthesis of phenolic compounds in tomato [21], and increased the expression of genes encoding epi-aristolochene synthase and cinnamoyl-CoA reductase [26].

Volatiles (VOCs) are a group of plant terpenes which play an important role in plant–pathogen interactions [27]. In tomato, BFO is able to induce dissemination of VOCs into the environment, such as (E)-2-hexenal, nonenal, (+)-2-carene, guaiacol, methyl salicylate, benzyl alcohol, and eugenol. Especially (E)-2-hexenal and methyl salicylate increased gradually after BFO treatment. (E)-2-hexenal increased by 92 % compared to the control [24]. Meanwhile, fifteen new compounds were detected in response to BFO. In addition, the synthesis and release of VOCs' quantity and quality were attenuated or delayed in young tomato leaves. The amount of monoterpene and sesquiterpenoids was markedly enhanced after BFO treatment in old leaves, but not significantly enough in young leaves [28].

Effect on Postharvest Fruits

Although BFO has no known antimicrobial activities, BFO has been confirmed to provide a broad-spectrum protection against decay of postharvest fruits caused by pathogen infection. The treatment with BFO reduced *B. cinerea* infection in grapes, kiwifruit and tomato, *Penicillium expansum* infection in apples, *Penicillium italicum* infection in citrus, *Colletotrichum musae* infection in bananas, and effectively inhibited natural postharvest diseases (Fig. 5.2) [21, 29]. The study has shown that BFO treatment could enhance control by

Fig. 5.2 Photo of grape clusters 10 days post-inoculation (CK, distilled water; BFO, burdock fructooligosaccharide) (Sun et al. 2013) [29]

Rhodotorula mucilaginosa against *Rhizopus* and blue mold decay of peaches, increasing chitinase and β-1,3-glucanase activity of *R. mucilaginosa* to reduce the natural decay of peaches [30]. In addition, BFO could reduce respiration rate, weight loss ,and titratable acidity to preserve the fruit quality and prolong the shelf life of postharvest grapes [29].

Mechanism of BFO in Induced Resistance

Induced Stomatal Closure

Pathogen entry into host tissue is a critical step in causing infection in plants. Stomata as natural surface openings are important entry sites of bacteria. Recent studies have shown that stomata are a checkpoint of host immunity and pathogen virulence in plants [31]. Elicitors such as chitosan, oligogalacturonic acid, and oligochitosan can also induce stomatal closure [32, 33].

Guo et al. [34] found that BFO could induce stomatal closure in *Pisum sativum*. The average width of stomatal apertures was decreased by 54% at a concentration of 1 mg/ml BFO, which was in a time- and concentration-dependent manner. BFO-induced stomatal closure is mediated by ROS and ROS-dependent NO production, as an effective barrier against pathogens [35, 36].

Systemic Acquired Resistance

Plants have developed multiple layers of defense mechanism against pathogen attack. Within minutes, pathogen recognition by the host triggers a variety of early responses, such as ion fluxes across membranes, pH changes, and reaction oxygen species (ROS) [37, 38]. Within hours, these events are followed by the induction of signaling secondary metabolites, such as salicylic acid (SA), jasmonates (JA), ethylene (ET), activation of defense-related genes and accumulation of pathogenesis-related proteins (PRs) [39, 40]. Systemic acquired resistance (SAR) is an inducible plant defense response that is triggered in many plants following infection by pathogens, dependent on SA and associated with PRs. Induced systemic resistance (ISR) in contrast to SAR is dependent on JA and ET [41]. As a natural elicitor, BFO can trigger pH changes, ROS, and nitric oxide in tobacco [21, 42]. Then BFO can induce disease resistance response through a salicylic acid-dependent signal pathway. BFO treatment enhanced the endogenous SA content and the levels of transcription of PR genes [21, 22].

Transcriptome Profile

The transcriptome is the complete set of transcripts in a cell for a specific developmental stage or mechanism. Various technologies have been developed to deduce the transcriptome, including gene microarray and RNA sequencing [43]. RNA sequencing has successfully been used in various plants. Wu et al. [44] identified numerous changes in the gene expression of floral sex determination in cucumbers by Solexa sequencing. Most of the genes which play crucial roles in the biosynthesis of active ingredient in *Salvia miltiorrhiza* were found by Solexa sequencing [45].

Guo et al. [26] compared differential expression profiling of tobacco after treatment with BFO or distilled water. Transcriptome profile analysis indicated that 412 genes expressed differently after BFO treatment (Fig. 5.3). Among them, expression of more than 30 genes associated with SA-mediated SAR was increased. Expression of other hormone related genes was changed by BFO treatment, such as JA, ET, abscisic acid, and gibberellic acid. The differentially expressed genes were mainly involved in stress responses, defense responses, biosynthetic processes, hormone responses, RNA biosynthetic processes, signaling pathways, and other processes. In addition, secondary metabolites such as total phenolics, flavonoids, lignin, and VOCs increased after BFO treatment, enhancing the resistance to pathogens and stress tolerance in plants and fruits.

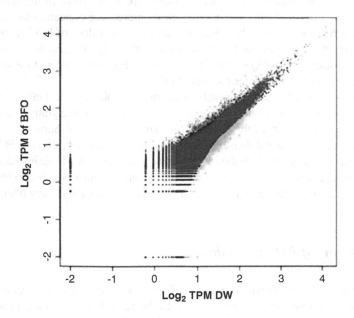

Fig. 5.3 Gene expression levels in tobacco after BFO and water treatment (Guo et al. 2012) [26]

In conclusion, the resistance induced by BFO in tobacco was linked with the differential expression of genes involved in plant hormone signaling pathways, especially salicylic acid-mediated pathway, and biosynthesis of secondary metabolites [26].

Applications and Future Trends of FOS

FOS have a number of interesting functional properties that make them important food ingredients. The nutritional and health benefits have been illuminated by many reviews in the recent years [46, 47]. Applications of FOS include the following:

Applications of FOS in Food

Presently, the concepts focus on the use of foods that promote a state of well-being, better health, and reduction of disease risk. Thus, recently a lot of attention has been paid to FOS present in diet. FOS as an effective prebiotic were demonstrated through both in vivo and in vitro assessments. FOS increased the number of *bifido-bacteria* and mostly decreased clostridia, resulting in a healthy gut environment.

Studies were carried out on two commercial strains of *Bifidobacterium* spp. cultured anaerobically in reconstituted nonfat dry milk (NDM) containing 0, 0.5 FOS. Growth promotion, enhancement of activity, and retention of viability of the cultures were enhanced when *Bifidobacterium* spp. were grown in the presence of FOS, and the effects of FOS increased with its increasing concentration and were maximal at 5 % (w/v) [48]. The effect of ingesting a low dose of FOS (5 g/day) by healthy human subjects on the fecal microflora especially *bifidobacteria* was investigated. Consumption of FOS (5 g per day) for 11 days resulted in close to one log cycle increase in *bifidobacteria* numbers [49].

Dietary fiber benefits health through a wide range of physiological effects. FOS are storage carbohydrates in a number of vegetables, fruits, and whole grains, which resist digestion and absorption in the stomach and small intestine of humans. Studies in patients with a conventional ileostomy have shown that mean excretion of FOS at the end of ileum was about 90 % of the ingested dose. They enter the large intestine where they will be available for fermentation, as demonstrated by increased breath hydrogen. Therefore, FOS have been found to fit well within the concept of dietary fiber [50].

Applications of FOS in Medicine

Human gut is colonized with a myriad of viruses, eukaryotes, and bacteria, and some diseases might result from dysbiosis. In the gastrointestinal tract, an adequate inflammatory response is critical to clear pathogenic bacteria. In an in vitro study,

FOS derived from *Arctium lappa* intriguingly alleviated inflammatory response on RAW264.7 cells by attenuating activation of NF-κB and MAPK pathway induced by LPS [51]. These results provided new insights into the underlying mechanism involved in the anti-inflammatory activity of FOS.

Numerous studies have reported the improvement of glucose homeostasis and lipid homeostasis by FOS feeding in a number of animal models, such as high-fat-diet-fed animals, genetically obese or diabetic mice, and streptozotocin-induced diabetic rats [52–54]. However, human studies have yielded inconsistent or negative results on the metabolism regulation by FOS. In a double-blind crossover design, intake of 20 g FOS/d for 4 weeks decreased basal hepatic glucose production, but had no detectable effect on insulin-stimulated glucose metabolism in healthy subjects [55]. In patients with nonalcoholic steatohepatitis, compared to placebo (maltodextrine), daily ingestion of 16 g/day FOS significantly decreased serum aminotransferases, aspartate aminotransferase after 8 weeks, and insulin level after 4 weeks, but this could not be related to significant effect on plasma lipids [56].

Accumulating evidence indicates that the symbiotic gut microbiota is instrumental in the control of host energy metabolism [57–59]. In addition, FOS improved glucose tolerance, increased enteroendocrine L-cell number and associated parameters, and reduced fat-mass development, oxidative stress, and low-grade inflammation. These results unequivocally imply that FOS could be used to control obesity and related metabolic disorders.

Irritable bowel syndrome (IBS) is a complex disorder characterized by chronic abdominal pain, discomfort, bloating, and alteration of bowel habits, in the absence of obvious organic abnormalities [60]. Some studies have investigated the effect of FOS in patients with IBS. Treatment with FOS led to a significant decrease in the intensity of digestive disorder as compared to the placebo product. Improvement in digestive comfort and in daily activities was also observed [61].

Studies with inulin and FOS have shown reduction of chemically induced aberrant crypts and prevention of colon cancer. According to Pool-Zobel et al. [62], a prebiotic effect resulting in the proliferation of *bifidobacteria* as well as of other bacteria could be responsible for the observed anticancer effects in rats. Dietary treatment with inulin/FOS (15%) incorporated in the basal diets for experimental animals resulted in reduction of the incidence of mammary tumors, the growth of transplantable malignant tumors, and the incidence of lung metastases of a malignant tumor in mice. It is reported that the dietary treatment with FOS/inulin significantly potentiated the effects of subtherapeutic doses of six different cytotoxic drugs commonly utilized in human cancer treatment [63].

Applications of FOS in Agriculture and Postharvest Fruits

FOS, in particular BFO, are a promising elicitor of disease control in plants and postharvest fruits. It is reported that BFO can increase the resistance of tomatoes to *B. cinerea* [24], that of cucumbers to *Colletotrichum orbiculare* [22], and that of tobacco to the tobacco mosaic virus [21]. Compared to elicitors, chemical fungicides are limited by the obvious problems of pollution and difficult degradation. Application of elicitors to induce natural disease resistance against postharvest diseases could be the focus [64]. In addition, the extraction of most FOS is very easy and effective. For example, a lot of BFO crude extracts are obtained by water extraction alone. The applications of induced resistance of FOS in agriculture will improve economic efficiency.

In postharvest disease control, the influence of BFO in controlling postharvest decay of fruits was notable, among tomato fruits, peaches, apple, banana, kiwi, citrus, and Kyoho grapes. The use of BFO to control postharvest disease has attracted much attention due to imminent problems associated with chemical agents, which easily include development of public resistance to fungicide-treated produce. As an edible material, FOS are not only nontoxic and biocompatible, but also have versatile functional properties, which agree with the demands of modern society. The applications of FOS should be further developed on induced resistance in the future.

Conclusion

In summary, FOS have a wide range of biological activity. On the one hand, it has emerged as one of the important candidates in the functional food market to promote a state of well-being. On the other hand, it has been demonstrated that it could reduce the risk of diseases. But there is a need for further research on the effect of FOS in plants, which will promote the applications of agriculture and postharvest disease control of BFO as a potent elicitor.

References

1. Campbell JM, Fahey Jr GC, Wolf BW. Selected indigestible oligosaccharides affect large bowel mass, cecal and fecal short-chain fatty acids, ph and microflora in rats. J Nutr. 1997;127(1):130–6.
2. Zentek J, Marquart B, Pietrzak T. Intestinal effects of mannanoligosaccharides, transgalactooligosaccharides, lactose and lactulose in dogs. J Nutr. 2002;132(6 Suppl 2):1682S–4.
3. Scholtens PA, Goossens DA, Staiano A. Stool characteristics of infants receiving short-chain galacto-oligosaccharides and long-chain fructo-oligosaccharides: a review. World J Gastroenterol. 2014;20(37):13446–52.
4. van de Wiele T, Boon N, Possemiers S, et al. Inulin-type fructans of longer degree of polymerization exert more pronounced in vitro prebiotic effects. J Appl Microbiol. 2007;102(2):452–60.
5. Kelly G. Inulin-type prebiotics—a review: Part 1. Altern Med Rev. 2008;13(4):315–29.
6. Benkeblia N. Fructooligosaccharides and fructans analysis in plants and food crops. J Chromatogr A. 2013;1313:54–61.
7. Der Agopian RG, Soares CA, Purgatto E, et al. Identification of fructooligosaccharides in different banana cultivars. J Agric Food Chem. 2008;56(9):3305–10.

8. Soleimani N, Hoseinifar SH, Merrifield DL, et al. Dietary supplementation of fructooligosaccharide (fos) improves the innate immune response, stress resistance, digestive enzyme activities and growth performance of Caspian roach (*Rutilus rutilus*) fry. Fish Shellfish Immunol. 2012;32(2):316–21.

9. Akrami R, Iri Y, Rostami HK, et al. Effect of dietary supplementation of fructooligosaccharide (fos) on growth performance, survival, lactobacillus bacterial population and hemato-immunological parameters of stellate sturgeon (*Acipenser stellatus*) juvenile. Fish Shellfish Immunol. 2013;35(4):1235–9.

10. Hermann M, Freire I, Pazos C. Compositional diversity of the yacon storage root. CIP Program Report, 425–432, 1999.

11. Vilhena SMC, Câmara FLA, Kakihara ST. O cultivo de yacon no Brasil. Hortic Bras. 2000;18(1):5–8.

12. Prata MB, Mussatto SI, Rodrigues LR, et al. Fructooligosaccharide production by penicillium expansum. Biotechnol Lett. 2010;32(6):837–40.

13. Muir JG, Shepherd SJ, Rosella O, et al. Fructan and free fructose content of common Australian vegetables and fruit. J Agric Food Chem. 2007;55(16):6619–27.

14. Fontana JD, Grzybowski A, Tiboni M, et al. Fructo-oligosaccharide production from inulin through partial citric or phosphoric acid hydrolyses. J Med Food. 2011;14(11):1425–30.

15. Niness KR. Inulin and oligofructose: what are they? J Nutr. 1999;129(7 Suppl):1402S–6.

16. Kim SS, Kim YJ, Rhee IK. Purification and characterization of a novel extracellular protease from *Bacillus cereus* KCTC 3674. Arch Microbiol. 2001;175(6):458–61.

17. Sangeetha PT, Ramesh MN, Prapulla SG. Recent trends in the microbial production, analysis and application of fructooligosaccharides. Trends Food Sci Technol. 2005;16(10):442–57.

18. Guio F, Rodríguez MA, Alméciga-Diaz CJ, et al. Recent trends in fructooligosaccharides production. Recent Pat Food Nutr Agric. 2009;1:221 30.

19. Saavedra JM, Tschornia A. Human studies with probiotics and prebiotics: clinical implications. Br J Nutr. 2002;87:S241–6.

20. Sauer J, Richter K, Pool-Zobel B. Products formed during fermentation of the prebiotic inulin with human gut flora enhance expression of biotransformation genes in human primary colon cells. Br J Nutr. 2007;97(5):928–38.

21. Wang F, Feng G, Chen K. Defense responses of harvested tomato fruit to burdock fructooligosaccharide, a novel potential elicitor. Postharvest Biol Tec. 2009;52(1):110–6.

22. Zhang PY, Wang JC, Liu SH, et al. A novel burdock fructooligosaccharide induces changes in the production of salicylates, activates defence enzymes and induces systemic acquired resistance to *Colletotrichum orbiculare* in cucumber seedlings. J Phytopathol. 2009;157(4):201–7.

23. Hao L, Chen K, Li G. Physiological effects of burdock oligosaccharide on growth promotion and chilling resistance of cucumber seedlings. J Shanghai Jiaotong University (agricultural Science) 2006, 24(1):6–12.

24. He PQ, Tian L, Chen KS, et al. Induction of volatile organic compounds of Lycopersicon esculentum Mill. and its resistance to Botrytis cinerea Pers. by burdock oligosaccharide. J Integr Plant Biol. 2006;48(5):550–7.

25. Dixon RA. Natural products and plant disease resistance. Nature. 2001;411(6839):843–7.

26. Guo M, Chen K, Zhang P. Transcriptome profile analysis of resistance induced by burdock fructooligosaccharide in tobacco. J Plant Physiol. 2012;169(15):1511–9.

27. Zhao N, Zhuang X, Shrivastava G, et al. Analysis of insect-induced volatiles from rice. Methods Mol Biol. 2013;956:201–8.

28. Zhang PY, Chen KS, He PQ, et al. Effects of crop development on the emission of volatiles in leaves of *Lycopersicon esculentum* and its inhibitory activity to *Botrytis cinerea* and *Fusarium oxysporum*. J Integr Plant Biol. 2008;50(1):84–91.

29. Sun F, Zhang P, Guo M, et al. Burdock fructooligosaccharide induces fungal resistance in postharvest Kyoho grapes by activating the salicylic acid-dependent pathway and inhibiting browning. Food Chem. 2013;138(1):539–46.

30. Zhang H, Liu Z, Xu B, et al. Burdock fructooligosaccharide enhances biocontrol of *Rhodotorula mucilaginosa* to postharvest decay of peaches. Carbohydr Polym. 2013;98(1):366–71.
31. Zeng W, Melotto M, He SY. Plant stomata: a checkpoint of host immunity and pathogen virulence. Curr Opin Biotechnol. 2010;21(5):599–603.
32. Lee S, Choi H, Suh S, et al. Oligogalacturonic acid and chitosan reduce stomatal aperture by inducing the evolution of reactive oxygen species from guard cells of tomato and Commelina communis. Plant Physiol. 1999;121(1):147–52.
33. Li Y, Yin H, Wang Q, et al. Oligochitosan induced *Brassica napus L.* production of NO and H_2O_2 and their physiological function. Carbohydr Polym. 2009;75(4):612–7.
34. Guo Y, Guo M, Zhao W, et al. Burdock fructooligosaccharide induces stomatal closure in *Pisum sativum*. Carbohydr Polym. 2013;97(2):731–5.
35. Khokon MA, Uraji M, Munemasa S, et al. Chitosan-induced stomatal closure accompanied by peroxidase-mediated reactive oxygen species production in Arabidopsis. Bio-sci Biotechnol Biochem. 2010;74:2313–5.
36. Melotto M, Underwood W, Koczan J, et al. Plant stomata function in innate immunity against bacterial invasion. Cell. 2006;126:969–80.
37. Grant M, Mansfield J. Early events in host-pathogen interactions. Curr Opin Plant Biol. 1999;2:312–9.
38. Nurnberger T, Scheel D. Signal transmission in the plant immune response. Trends Plant Sci. 2001;6:372–9.
39. Kombrink E, Somssich IE. Defence responses of plants to pathogens. Adv Bot Res. 1995;21:1–34.
40. Blumwald E, Aharon GS, Lam BC-H. Early signal transduction pathways in plant-pathogen interactions. Trends Plant Sci. 1998;3(9):342–6.
41. Vallad GE, Goodman RM. Systemic acquired resistance and induced systemic resistance in conventional agriculture. Crop Sci. 2004;44:1920–34.
42. Guo M, Chen K. Transcriptome profile analysis and signal transduction of resistance induced by burdock fructooligosaccharide in tobacco. Jinan: Shandong University; 2014.
43. Hoen PAC, Ariyurek Y, Thygesen HH, et al. Deep sequencing-based expression analysis shows major advances in robustness, resolution and interlab portability over five microarray platforms. Nucleic Acids Res. 2008;36, e141.
44. Wu T, Qin Z, Zhou X, et al. Transcriptome profile analysis of floral sex determination in cucumber. J Plant Physiol. 2010;167(11):905–13.
45. Hua WP, Zhang Y, Jie S, et al. De novo transcriptome sequencing in salvia miltiorrhiza to identify genes involved in the biosynthesis of active ingredients. Genomics. 2011;98(4):272–9.
46. Flamm G, Glinsmann W, Kritchevsky D, et al. Inulin and oligofructose as dietary fiber: a review of the evidence. CRC Critical Rev Food Sci Nutr. 2001;41:353–62.
47. Flickinger EA, Loo JV, Fahey GC. Nutritional responses to the presence of inulin and oligofructose in the diets of domesticated animals. CRC Critical Rev Food Sci Nutr. 2003;43:19–60.
48. Shin HS, Lee JH, Pestka JJ, et al. Growth and viability of commercial Bifidobacterium spp in skim milk containing oligosaccharides and Inulin. J Food Sci. 2000;65(5):884–7.
49. Rao V. The prebiotic properties of oligofructose at low intake level. Nutr Res. 2001;21:843–8.
50. Cherbut C, Michel C, Lecannu G. The prebiotic characteristics of fructooligosaccharides are necessary for reduction of TNBS-induced colitis in rats. J Nutr. 2003;133(1):21–7.
51. Liu J, Pan X, Song Z, et al. Anti-inflammatory effect of burdock fructo-oligosaccharide on lipopolysaccharide-stimulated RAW264.7 cell. J Shandong Univ (Health Sci). 2012;50(12):41–6.
52. Kok NN, Taper HS, Delzenne NM. Oligofructose modulates lipid metabolism alterations induced by a fat-rich diet in rats. J Appl Toxicol. 1998;18(1):47–53.
53. Cani PD, Neyrinck AM, Maton N, et al. Oligofructose promotes satiety in rats fed a high-fat diet: involvement of glucagon-like Peptide-1. Obes Res. 2005;13(6):1000–7.
54. Delmee E, Cani PD, Gual G, et al. Relation between colonic proglucagon expression and metabolic response to oligofructose in high fat diet-fed mice. Life Sci. 2006;79(10):1007–13.

55. Luo J, Rizkalla SW, Alamowitch C, et al. Chronic consumption of short-chain fructooligosac-charides by healthy subjects decreased basal hepatic glucose production but had no effect on insulin-stimulated glucose metabolism. Am J Clin Nutr. 1996;63(6):939–45.
56. Daubioul CA, Horsmans Y, Lambert P, et al. Effects of oligofructose on glucose and lipid metabolism in patients with nonalcoholic steatohepatitis: results of a pilot study. Eur J Clin Nutr. 2005;59(5):723–6.
57. Holmes E, Li JV, Marchesi JR, et al. Gut microbiota composition and activity in relation to host metabolic phenotype and disease risk. Cell Metab. 2012;16(5):559–64.
58. Suzuki TA, Worobey M. Geographical variation of human gut microbial composition. Biol Lett. 2014;10(2):20131037.
59. Fukuda S, Ohno H. Gut microbiome and metabolic diseases. Semin Immunopathol. 2014;36(1):103–14.
60. Spiller R, Aziz Q, Creed F, et al. Guidelines on the irritable bowel syndrome: mechanisms and practical management. Gut. 2007;56(12):1770–98.
61. Paineau D, Payen F, Panserieu S, et al. The effects of regular consumption of short-chain fructo-oligosaccharides on digestive comfort of subjects with minor functional bowel disorders. Br J Nutr. 2008;99(2):311–8.
62. Pool-Zobel BL, Van Loo J, Rowland IR, et al. Experimental evidences on the potential of prebiotic fructans to reduce the risk of colon cancer. Br J Nutr. 2002;87:S273–81.
63. Taper HS, Roberfroid MB. Inulin/oligofructose and anticancer therapy. Br J Nutr. 2002;87:S283–6.
64. Terry LA, Joyce DC. Elicitors of induced disease resistance in postharvest horticultural crops: a brief review. Postharvest Biol Tec. 2004;32(1):1–13.

Chapter 6
Chitosan-Elicited Plant Innate Immunity: Focus on Antiviral Activity

Marcello Iriti and Elena Maria Varoni

Abstract Immunity represents a trait common to all living organisms, and animals and plants share some similarities. Therefore, in susceptible host plants, a complex defence machinery may be stimulated by elicitors. Among these, chitosan deserves particular attention because of its proved efficacy. This survey deals with the antiviral activity of chitosan, focusing on its perception by the plant cell and mechanism of action. Emphasis has been paid to benefits and limitations of this strategy in crop protection, as well as to the potential of chitosan as a promising agent in virus disease control.

Keywords Plant activators • Plant innate immunity • PAMPs/MAMPs • PRRs • Viral diseases • Plant viruses • Crop protection • Environmental safety

Virus Biology, Infection Process and Disease Control

A virus is a nucleoprotein consisting of nucleic acid (RNA or DNA) and protein, the latter forming the capsid, a protective coat around the former. This nucleoprotein multiplies only in living cells and may cause a multitude of diseases in all organisms. Some viruses attack humans, animals or both, causing diseases such as influenza, polio, rabies and acquired immunodeficiency syndrome (AIDS), others infect higher plants and others microorganisms as well, such as fungi and bacteria. The total number of viruses known to date exceeds 2000, and nearly half of these has the ability to cause diseases in plants [1].

The nucleic acid of most plant viruses consists of RNA protected by the protein shell, also involved in vector transmissibility (mainly by arthropods), infectivity

M. Iriti, Ph.D. (✉)
Department of Agricultural and Environmental Sciences, Milan State University, Milan, Italy
e-mail: marcello.iriti@unimi.it

E.M. Varoni, Ph.D.
Department of Biomedical, Surgical and Dental Sciences, Dental Unit II, San Paolo Hospital, Milan State University, Milan, Italy

© Springer Science+Business Media New York 2016   65
H. Yin, Y. Du (eds.), *Research Progress in Oligosaccharins*,
DOI 10.1007/978-1-4939-3518-5_6

and symptom development. Plant viruses enter cells only through wounds, or by vectors, or by deposition into a ovule by an infected pollen grain. They are transmitted either in a persistent or non-persistent manner by insects (mostly aphids). Persistent transmission means that once the insect vector becomes infectious, it remains in this condition for the rest of its life cycle. Non-persistent transmission implies that the insect can acquire the virus by a brief probe on an infected plant, afterwards it can transmit the virus directly: the vector will lose viral particles after probing only one or two times on healthy plants. The mode of transmission has implications on the way a virus develops in the field and its management. Once inside, they move from one cell to another through the plasmodesmata connecting adjacent cells, multiplying in them. Then, viruses reach the phloem, spread systematically throughout the plant and, finally, re-enter the parenchyma cells adjacent to the phloem through the plasmodesmata. Therefore, local and/or systemic symptoms may arise and result from cell-to-cell or phloematic translocation, respectively [2, 3]. In general, virus infections are sporadic and their levels depend heavily on seasonal conditions, differing greatly in the years and according to locations. Early infections can lead to stunting, reduced tillering and plant death, and losses can be high. Late infections have less impacts, but can still affect seed quality [2, 3].

Plant viruses differ from fungal pathogens since no curative treatment (fungicides) is available and, therefore, viral disease control aims at prevention through integrated management practices to control the virus source, aphid populations and virus transmission into crops. Alternatively, the use of virus-resistant cultivars and stimulation of the plant's own defence mechanisms by elicitors may represent, in some cases, two effective strategies [4].

Plant Innate Immunity

Disease is a rare outcome in the spectrum of plant–microbe interactions and plants have (co)evolved a complex set of defence mechanisms to hinder pathogen challenging and, in most cases, prevent infection. The battery of defence reactions includes physical and chemical barriers, both preformed (or constitutive or passive) and inducible (or active), depending on whether they are pre-existing features of the plant or are switched on after challenging (Table 6.1). When a pathogen is able to overcome these defences, disease ceases to be the exception. Three main explanations support this rule: (1) plant is not a substrate for microbial growth and does not support the lifestyle of the invading pathogen; (2) constitutive barriers prevent colonisation of plant by pathogen; (3) plant recognises pathogen, by its innate immune system, and then activates inducible defences [5].

The host ability of responding to an infection is determined by genetic traits of both the plant itself and the pathogen. Some resistance mechanisms are specific for plant cultivars and certain pathogen strains. In these cases, plant resistance (R) genes, encoding for receptors, recognise pathogen-derived molecules (specific elicitors) resulting from the expression of avirulence (*avr*) genes (Table 6.2). This

Table 6.1 Plant defence mechanisms

	Structural	Chemical
Constitutive (passive, preformed)	Anatomical barriers (trichomes, cuticle, cell wall)	Preformed inhibitors (phytoanticipins: glycosides, saponins, alkaloids), antifungal proteins (lectins) and ribosome inactivating proteins (RIP)
Inducible (active)	Cell wall strengthening (callose, lignin and suberin appositions; oxidative extensin cross-linking)	Oxidative burst, hypersensitive response (HR), phytoalexins (phenylpropanoids), pathogenesis-related proteins

Table 6.2 Plant innate immunity

Type of resistance	Elicitors
Host (specific) resistance	Specific elicitors, encoded by the *avr* genes of certain pathogen strains (gene-for-gene theory)
Non-host (basal) resistance	General exogenous (race-non-specific MAMPs[a]) and endogenous (plant-derived oligogalacturonides) elicitors

[a]*MAMPs* microbe-associated molecular patterns

gene-for-gene relation, also known as host resistance, triggers inducible barriers, i.e, a cascade of events leading to systemic acquired resistance (SAR). In addition, another type of resistance is activated through the recognition, by plant receptors, of general (race-nonspecific) elicitors, pathogen- or microbe-associated molecular patterns (PAMPs or MAMPs) including mainly lipopolysaccharides, peptidoglycans, flagellin, fungal cell wall fragments, lipid derivatives (sterols and fatty acids), proteins, double stranded RNA and methylated DNA (Table 6.2). This non-host or basal resistance can also be induced by endogenous, plant-derived, general elicitors (host- or damage-associated molecular patterns, HAMPs or DAMPs), such as oligogalacturonides released from the plant cell wall by fungal hydrolytic enzymes (Table 6.2) [6–8]. In any case, the spectrum of defence reactions triggered by both types of resistance, that collectively represent the plant innate immune system, is rather similar [9]. Immunity may be expressed locally (local acquired resistance, LAR), in the infected cells, or in uninfected distal tissues (SAR), probably because of one or more endogenous systemically translocated (or volatile) signals that activate defence mechanisms in plant organs distal from the initial site of infection [10].

Recognition of a biotic stress by cell entails physical interaction of a stimulus (elicitor) with a receptor. According to the receptor/ligand model, the constitutively expressed plant R genes encode proteins that possess domains characteristic of typical receptors responsible for the innate immunity in mammals and *Drosophila*. These proteins, also known as pattern-recognition receptors (PRRs), can be grouped into different classes according to certain common structural motifs. Many R proteins contain a leucine-rich repeat (LRR) domain involved in recognition specificity. Some receptor-like proteins (RLPs) possess an extracellular LRR anchored to a trans membrane domain, whereas other R genes encode receptor-like kinases

(RLKs) with extracellular LRR and a cytoplasmic kinase domain. Another group of receptors possesses amino acid sequences with strong similarity to nucleotide binding (NB) sites. These NB-LRR proteins are likely localised in cytoplasm, with a putative coiled-coil (CC) domain or a region with similarity to the Toll and interleukin-1 receptor (TIR) at the N terminus. In particular, CC (in CC-NB-LRR) can be a leucine zipper (LZ), whereas TIR (in TIR-NB-LRR) plays a central role in the immune and inflammatory responses of mammals and *Drosophila*. Finally, other receptor proteins contain one or two intracellular serine–threonine kinases [6–8].

Among MAMPs, chitosan, a deacetylated chitin derivative, is worthy of special attention because of its use in chemical-induced resistance and efficacy against virus diseases [11]. Like a general elicitor, CHT is able to prime an aspecific, long-lasting and systemic immunity (SAR), possibly by binding to a receptor in the plant cell surface [11–16].

Chitosan Chemistry

Chitosan is a linear, polycationic heteropolysaccharide discovered by Rouget in 1859. It consists of two monosaccharides, N-acetyl-D-glucosamine (2-acetamido-2-deoxy-β-D-glucopyranose, $C_8H_{15}NO_6$, MW=221.2), the repeat unit of insoluble chitin, and D-glucosamine (2-amino-2-deoxy-β-D-glucopyranose, $C_6H_{13}NO_5$, MW = 179.17) linked together by β-$(1 \rightarrow 4)$ glycosidic bonds (Fig. 6.1). Chitosan is low-cost produced by exhaustive alkaline or enzymatic deacetylation from chitin, widely distributed in nature, mainly as the structural component of the arthropod exoskeleton (crustaceans and insects) and fungal cell wall. Therefore, the relative amount of the two monosaccharides in this partially N-deacetylated chitin derivative may vary, with samples of different deacetylation degree (DD), molecular weight (MW), polymerization degrees (PD), viscosities and pK_a values, and the

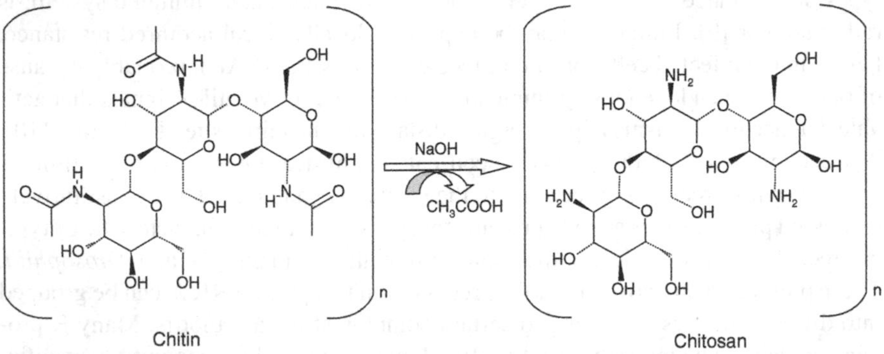

Chitin   Chitosan

Fig. 6.1 Chitin N-deacetylation through alkaline hydrolysis: acetamido [-NHCOCH$_3$] groups are deacylated in chitosan molecule

term "chitosan" does not refer to a uniquely defined compound, but describes a heterogeneous family of copolymers, commercially available form a number of suppliers. Noteworthy, all these chemical properties greatly affect the chitosan physicochemical characteristics, which, in turn, govern almost all its biological applications.

In nature, chitosan is also found in fungal cell wall, even if it differs from invertebrate chitosan: whereas the acetyl groups in chitosan produced from crustacean chitin are uniformly distributed along the polymer chain, a chitosan of similar DD isolated from fungal cell wall would possess acetyl residues that are grouped into clusters [17].

Chitosan also possesses several favourable biological properties, above all biodegradability and low toxicity. It is susceptible to degradation by both specific and non-specific enzymes, including chitinases, chitosanases, lysozymes, cellulases, hemicellulases, proteases, lipases and glucanases, and its low toxicity has been also documented in human studies [18].

Chitosan Antiviral Activity

Chitosan possesses a well documented, broad spectrum, direct, antimicrobial activity against filamentous fungi, yeasts, Gram positive and Gram-negative bacteria [19–21]. Conversely, its activity against plant viruses is due to the capability of the polymer to stimulate the plant immune response [11]. In this regards, a direct antiviral activity of chitosan was ruled out in TNV-inoculated bean leaves, and the protective effects of treatment were attributed to the elicitation of the plant defence mechanisms [22].

The elicitor activity of chitosan was first demonstrated in the interaction between pea (*Pisum sativum*) and the fungal pathogen *Fusarium solani* [23]. Similarly, the capacity of inducing resistance against viral diseases by chitosan is known since years. Initially, it was shown that treatment of bean (*Phaseolus vulgaris*) leaves with the polymer decreased the number of local necrotic lesions caused by alfalfa mosaic virus (AFM) by triggering SAR [24, 25]. In particular, treatment of the lower surface of a leaf stimulated resistance in its upper surface, treatment of one half of a leaf induced resistance on the other untreated half, and treatment of lower leaves elicited defence mechanisms in upper leaves [24, 25]. This finding prompted further studies aimed at demonstrating that chitosan is able to induce resistance against a number of local and systemic viral infections in different host plants belonging to different botanical families [12]. Noteworthy, chitosan prevented infection with viruses characterised by different structures and genome expression mechanisms, thus suggesting that it may suppress infection irrespective of the type of virus, by stimulating the plant's own defence machinery [12]. The polymer also inhibited infection with potato spindle tuber viroid (PSTV), when sprayed on tomato leaf prior or not later 1–3 h after inoculation or added to the inoculum [26] (*Solanum lycopersicum*).

Even if viral infections can not be directly controlled by conventional agrochemicals, unfortunately, the reliable use of chitosan in crop protection is still hampered by a series of intrinsic and extrinsic factors. The formers are related to the physico-chemical properties of the polysaccharide, whereas the latter are due to the genetic traits of both host plant and pathogen as well as to environmental factors.

Factors Affecting Antiviral Activity of Chitosan

Molecular weight of chitosan greatly affects its ability to suppress viral infections by eliciting the host defence response. In bean plants, the degree of chitosan-induced resistance to the systemic pathogen bean mild mosaic virus (BMMV) increased as its MW decreased [27]. After chitosan depolymerisation obtained by enzymatic hydrolysis, the fractions with the lowest MW, 2.2 and 1.2 kDa, exhibited a higher antiviral activity than that of the fractions with 10.1, 30.3 and 40.4 kDa, whereas monomers glucosamine and N-acetylglucosamine did not show any activity. Intriguingly, with a DD of 85%, in samples with MW 1.2 and 2.2 kDa, on average one or two monomers are acetylated, possibly located at the end of the polymeric chain (at least in chitosan derived from crustacean chitin), thus suggesting a spatial interaction between particular chitosan conformations and putative receptor sites on plant cell [27]. These results were corroborated by Davydova and colleagues, who obtained chitosans with different MW and DD after depolymerisation by enzymatic and chemical hydrolysis. Chitosan derivatives from 17.0 to 2.0 kDa inhibited the formation of local necrotic lesions by systemic tobacco mosaic virus (TMV) in tobacco plant (*Nicotiana tabacum*) by 50–90%. These authors also demonstrated that the antiviral activity of their samples only marginally depended on DD [28]. Similarly, DD of krill and crab chitosans within the range of 60–98% caused no significant effect on its activity against AMV on bean plants [29]. These authors also reported that glucosamine exhibited a significant antiviral activity, and a considerably lower yet substantial activity was also shown for N-acetylglucosamine [29]. Enzymatic degradation of high MW-chitosans by fungal chitinases from *Aspergillus fumigatus* significantly increased their ability in suppressing local necrotic lesions caused by TMV inoculation on tobacco plants [30, 31]. In general, an increase of antiviral activity in low MW-chitosans may be due to their better penetrating ability across the leaf epidermal tissues. Noteworthy, stomatal uptake experiments carried out on bean leaves showed that chitosan entering stomata is determinant for the induction of resistance to TNV [22].

Despite these results, the data on the dependence of chitosan antiviral activity on it structure, mainly MW, are still inconsistent and controversial, since it has been reported that high-polymeric chitosans possess higher antiviral properties too. In potato (*Solanum tuberosum*) plants, 120-kDa chitosan obtained from krill was more effective than crab chitosans of 3 and 36 kDa against systemic infection of potato virus X (PVX), and similar result were documented by the same authors in bean plants inoculated with AMV [32, 33].

This contrasting results may be due to the different sources of the polymer, for instance crab or krill chitosan, and to the heterogeneity of derivatives (oligomers) arising from different depolymerisation processes. Depolymerisation can be obtained by many enzymes including chitinases, chitosanases, glucanases, cellulases, hemicellulases, pectinases, proteases and lipases, or by chemical hydrolysis for instance with hydrogen peroxide. However, independently from the polymer source, it has been suggested that, in contrast to chemical hydrolysis, enzymatic depolymerisation excludes the formation of bioactive oxidised groups in the oligosaccharides. Therefore, in the case of chemical hydrolysis, the higher antiviral activity of low MW chitosans is determined not so much by the change of the polysaccharide PD as by the appearance of oxidised groups in the oligomers [28, 34, 35]. At the same time, low MW chitosans obtained by chemical or enzymatic hydrolysis from the same source may differ in the distribution of acetate residues, which, in turn, may affect the antiviral properties of derivatives [35].

Though chitosan inhibited virus infection in plant species belonging to different botanical families, its activity was shown to be higher in Fabaceae than in other families [12, 24, 25], possibly because this activity seems to be mediated by the plant's own defence response. Generally, bean and pea plants were more responsive to treatment, acquiring high levels of long-lasting resistance to different viral or fungal infections after one application with low concentrations of chitosan [23–25]. Differently, Solanaceae members, such as tobacco, potato, tomato and stramony (*Datura stramonium*) plants proved to be more refractory even to repeated treatments with high chitosan concentrations [36–38]. On the other hand, chitosan was not able to elicit resistance to systemic infections with cauliflower mosaic virus (CaMV), turnip mosaic virus (TuMPV) or radish mosaic virus (RaMV) in cabbage (*Brassica campestris*, Cruciferae) [36].

Mechanisms of Chitosan Antiviral Activity

Sensing of Chitosan by Plant Cell

Perception of elicitors by cell represents the first step to trigger an effective plant immune response. As previously introduced, PRRs are able to recognise invariant and conserved structures usually originating from microbial surfaces, essential for microbial metabolism and not present in the host (with the exception of HAMPs/ DAMPs). Although the role of chitosan as SAR inducer has been convincingly demonstrated, little is known about the sensing of chitosan by plant cell. To date, only a putative receptor for chitosan has been described by Chen and Xu, who isolated a 78 kDa chitosan-binding protein from cabbage leaves [13]. In addition, in calli of *Cocos nucifera*, chitosan induced the expression of genes with high similarities to DNA sequences of RLKs [39].

For chitin, high-affinity binding sites were reported at the surface of suspension-cultured cells of different species: tomato [40], rice (*Oryza sativa*) [41], soybean (*Glycine max*) [42], wheat (*Triticum aestivum*), barley (*Hordeum vulgare*) and carrot (*Daucus carota*) [43]. More recently a plasma membrane receptor for chitin was identified and purified in rice cells, both at gene and protein levels [44]. This chitin elicitor-binding protein (CEBiP) structurally differs from the two major classes of PRRs in plants, RLKs and RLPs, both groups containing extracellular LRRs [44]. The mature glycoprotein CEBiP harbours two extracellular lysine motifs (LysMs) of approximately 40 amino acids and a transmembrane domain at the C-terminus, but it lacks any cytosolic domain for signal transduction, such as intracellular kinase domains normally present in RLKs, thus suggesting that additional factors for downstream chitin signalling through the plasma membrane into the cytoplasm are necessary [44]. Furthermore, knocking down *CEBiP* expression compromised the plant immune response [44]. LysM domains generally occur in a variety of peptido-glycan- and chitin-binding proteins, suggesting that they may be directly involved in glycan perception [45]. In *Arabidopsis thaliana*, a second chitin receptor was identified by the same authors, a chitin elicitor receptor-like kinase (CERK1) [46]. The KO mutants for *CERK1* completely lost the ability to activate an immune response, in particular mitogen-activated protein kinase (MAPK) activation, reactive oxygen species (ROS) generation and defence gene expression [46, 47]. Similarly to CEBiP, CERK 1 is a plasma membrane protein with three LysMs in the extracellular domain, but, differently from CEBiP, CEPK1 harbours an intracellular Ser/Thr kinase domain [46].

In any case, it is still under debate whether chitosan is perceived by chitin receptors. In in vitro studies, CERK1 ectodomain was found to bind chitin polymers with higher affinity than chitin oligomers (PD 4–8), but chitosan very weakly bound to this receptor [48, 49]. Differently form chitin, chitosan is a polycationic molecule able to interact with the outward-facing, negatively charged hydrophilic groups of phospholipid bilayers, and these perturbations may be sufficient to elicit the plant defence response [45].

Defence Reactions Triggered by Chitosan in Plant Cell

Early events following chitosan perception consist of changes in ion fluxes and plasma membrane depolarization. Transiently increased cytosolic Ca^{2+} concentration was demonstrated by Zuppini et al. [50] in soybean cells, with a maximal peak reached after about 3 min upon chitosan administration and falling back to the basal level after about 5 min. More recently, Amborabé and colleagues reported that plasma membrane depolarization peaked transiently 10–15 min after chitosan treatment in *Mimosa pudica* motor cells, with a correlated cytoplasm acidification and inhibition of plasma membrane H^+-ATPase activity [51]. Inhibitory effects on the proton-pumping activity of this enzyme also compromised the uptake of co-transported carbohydrates and amino acids as well as other H^+-mediated processes [51].

The activation of a Ca^{2+}-dependent callose synthase represents another rapid, effective and specific cell response to chitosan elicitation [52, 53], as shown in monocotyledonous and dicotyledonous species [54, 55]. Moreover, chitosan-induced callose apposition was mechanistically correlated to the antiviral activity of the polymer in different pathosystems, namely bean/tobacco necrosis virus (TNV), bean/tomato bushy stunt virus (TBSV) and tobacco/TNV [54, 56, 57]. In particular, the efficacy of chitosans with different MWs (6–735 kDa) to stimulate callose synthesis in bean leaf fragments was evaluated and correlated with their capability in inducing resistance to TNV [58]. Polymers with 76, 120 and 139 kDa were the most effective in stimulating callose apposition in comparison with those having lower or higher MW, and, interestingly, the intensity and pattern of callose deposition in leaf tissues positively correlated with chitosan-induced resistance to TNV, with the 76-kDa polymer reducing by 95 % the virus necrotic lesions [58]. The role of callose, a β-1,3-D-glucan, in limiting virus spreading is well established [59, 60]. Extracellular callose deposition, around plasmodesmata (callose collar), may constrain the cell-to-cell transport of viral particles; similarly, their long-distance transport, along the phloem vessels, may be restrained because of callose apposition in pores of the phloem sieves [61].

One of the largest families of Ser/Thr protein kinases is represented by MAPKs, and MAPK-mediated phosphorylation of transcription factors represents a key step in controlling defence gene expression [62]. Activation of MAPK cascade by chitosan was documented [39, 63]. With relation to antiviral activity, a novel chitosan-induced Ser/Thr protein kinase gene was isolated in tobacco plants and designated as oligochitosan-induced protein kinase (oipk) [64]. Antisense expression of oipk decreased phenylalanine ammonia-lyase (a key enzyme in phytoalexin biosynthesis) activity and resistance to TMV in transformed plants [64], and, more recently, in the same host, other authors reported a positive correlation between this transduction protein and TMV resistance, PAL and peroxidase activities and mRNA levels of PAL and two pathogenesis-related (PR) proteins, chitinase and β-1,3-glucanase [65].

The increased production of ROS by plant, including superoxide anion ($^{\cdot}O_2^{-}$), hydrogen peroxide (H_2O_2) and hydroxyl radical ($^{\cdot}OH$) is an important early-induced defence response to pathogen attack. Among these by-products of molecular oxygen, H_2O_2 plays a fundamental role due to its mobility (it possesses neither negative charges nor unpaired electrons), relatively low reactivity and broad-spectrum activity. In fact, it exerts a direct antimicrobial activity, besides being involved in cell wall strengthening by monolignol polymerization and lignin apposition, and oxidative cross-linking of hydroxyproline-rich glycoproteins (extensins), though these processes do not seem to be relevant in plant–virus interaction [58, 66, 67]. Hydrogen peroxide can also cross the plasma membrane and affect cell signalling, by interacting with reactive nitrogen species (RNS) [68]. Production of H_2O_2 and nitric oxide (NO) after chitosan treatment was observed in epidermal cells of tobacco, rapeseed (Brassica napus) and pea leaves [69–71], in pearl millet (Pennisetum glaucum) seedlings [72] and in rice cell suspensions [73].

A fine-tuned orchestration of molecular signals, particularly Ca^{2+}, H_2O_2 and NO is responsible for the onset of hypersensitive reaction (HR), a rapid and programmed death of plant cells (PCD) at sites of attempted pathogen penetration [68, 74]. This reaction may involve just a single cell (invisible HR) or extensive and visible tissue areas, and it is particularly effective against viruses, because these disease agents necessitate of the biosynthetic machinery of the healthy plant cell [4]. A chitosan-induced, calcium-mediated PCD, in soybean cells, was studied by Zuppini and co-workers [50]. They observed cytoplasm shrinkage, chromatin condensation and an increased activity of caspase-like proteases, morphological and biochemical features of PCD [50]. Similar morphological hallmarks were reported in tobacco BY2 cells, besides internucleosomal DNA fragmentation with a distinct DNA-laddering pattern (Fig. 6.2) [54]. Interestingly, cell death kinetic induced by chitosan was delayed by treatment with a calcium channel blocker, and HR was correlated to reduced TNV infection on tobacco plants [54]. On the other hand, a H_2O_2-independent form of PCD was demonstrated in tobacco cell cultures [75]. Cell death phenomena triggered by chitosan administration were also documented on epidermis of pea seedlings [76] and on sycamore (*Acer pseudoplatanus*) cultured cells [77].

The role of hormonal signalling and crosstalk in plant–virus interaction is somewhat controversial [22]. In general, SAR is associated with the accumulation of salicylic acid (SA) and PR proteins, and is dependent on the regulatory protein NPR1 (non-expressor of PR-1 gene). On the contrary, induced systemic resistance (ISR), which can be stimulated by beneficial rhizobacteria and the fungus *Trichoderma* spp. colonising the root system, does not require SA, can occur without the production of PR proteins, and is dependent on ethylene and jasmonic acid (JA) as well as NPR1. Finally, β-aminobutyric acid-induced resistance (BABA-IR) involves both SA- and abscisic acid (ABA)-dependent defence mechanisms, depending on the challenging pathogen [67]. The importance of jasmonic acid (JA) for signal transduction after elicitation of tomato plants with chitosan was demonstrated since years by Doares and colleagues [78]. Stimulation of octadecanoic pathway and accumulation of 12-oxo-phytodienoic acid (the precursor of JA) and JA were further documented in chitosan-treated rice leaves [79]. On the contrary, after chitosan administration, methyl-salicylic acid (the methyl ester of SA) accumulated in the same tissues of the same plant species [80], and SA biosynthesis increased in TMV-infected tobacco leaves [81]. In bean leaf tissues, it was shown that chitosan-induced callose apposition was regulated by ABA, whose accumulation was in turn correlated with resistance to TNV [82]. In fact, an inhibitor of ABA biosynthesis administered before chitosan treatment reduced both callose deposition and plant resistance to the virus, whereas exogenous ABA application induced a significant resistance to TNV, thus indicating the involvement of ABA in these processes [82]. Intriguingly, the transcription levels of *BnOIPK*, a MAPK gene cloned in *Brassica napus*, rapidly increased in leaf tissues after chitosan and JA treatments, but only slightly after SA and ABA exposure [83]. More recently, it was suggested that ethylene pathway does not seem to be involved in the antiviral activity of chitosan [22]. An ethylene-independent, chitosan-induced resistance was

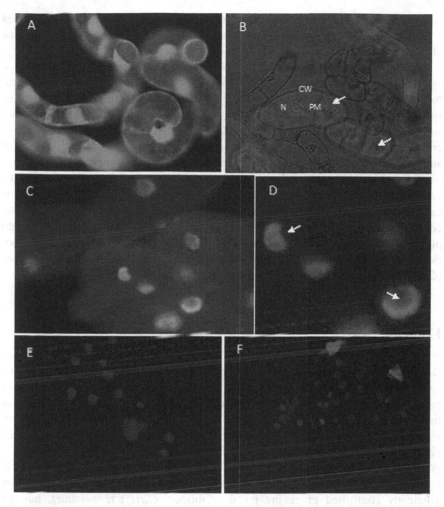

Fig. 6.2 Morphological features of programmed cell death (PCD) induced by 0.005 % chitosan treatment for 6 h in tobacco BY2 cells visualised by fluorescence microscope (Olympus BX50, Tokyo, Japan). (**a**) Viability of control cells treated with 0.05 % acetic acid (the solvent of chitosan) for 6 h and stained by fluorescein diacetate (FDA); bar 30 μm. (**b**) Viability and cytoplasm shrinkage (*arrows*) in chitosan-treated cells stained bt FDA; N, nucleus; CW, cell wall; PM, plasma membrane; bar 30 μm. (**c**) Chitosan-induced chromatin condensation; nuclei were stained by Hoechst 33258; bar 20 μm. (**d**) *Arrows* indicate chitosan-induced chromatin condensation; nuclei were stained by Hoechst 33258; bar 10 μm. (**e**) Terminal deoxynucleotidyl transferase dUTP nick end labelled (TUNEL) assay (In Situ Cell Death KIT TMR red, Roche, Basel, Switzerland); TMR red-negative control cells treated with 0.05 % acetic acid for 6 h; bar 40 μm. (**f**) Chitosan-treated cells containing DNA stand nicks characteristic of PCD detected by TUNEL assay (terminal deoxynucleotidyl transferase adds labelled dUTP to the 3′ OH ends of either single- or double-stranded DNA); TMR red-positive apoptotic nuclei co-stained with the DNA-binding dye DAPI; bar 40 μm

demonstrated by a pharmacological approach with inhibitors of the main enzymes of ethylene biosynthesis, a donor and a precursor of the hormone, and an inhibitor of its perception [22].

Finally, other events associated to SAR establishment, such as phytoalexin biosynthesis and PR protein translation, while effective against bacterial and fungal pathogens, do not seem to exert a significant role in preventing virus replication or spreading, even if the effects of chitosan on these late steps of SAR are known since decades in plant–fungus interaction [23, 84]. In particular, the role of antimicrobial secondary metabolites seems to be of little importance, at least in terms of antiviral defences, though their relevance against fungal and bacterial diseases has been demonstrated [67]. Similarly, among the PR proteins up to date isolated, antiviral agents are not included, with the exception of ribonuclease [4]. This enzyme can depolymerise viral RNA thus inhibiting its replication and translation, and chitosan-induced increase in ribonuclease activity was documented in potato plants and correlated with their resistance to PVX infection [33]. A 18 kDa ribonuclease (PR-10) able to degrade TMV RNA was also isolated in hot pepper (*Capsicum annuum*) [85].

Epilogue

Despite the benefits of chitosan in crop protection and its limitations in plant disease control were previously illustrated in this survey, emphasis should also be paid to some "positive side-effects" which may derive from the polymer application [86]. These include the possibility of rising the levels of bioactive secondary metabolites in plant and produce healthier foods, as shown in bean and grapevine [22, 87, 88]. In particular, chitosan increased polyphenol and melatonin contents in grapes and the corresponding experimental wines, as well as their antiradical activity [87, 88]. Interestingly, the new chitosan formulation used in one of this study efficiently controlled grapevine powdery mildew (*Erisyphe necator*), but not downy mildew (*Plasmopara viticola*) [87]. The polymer was also effective in reducing mycotoxin contamination of fungus-infected cereal grains [89]. In addition, in open field trials on bean plants, chitosan treatments did not incur fitness costs, a potential detrimental trait of induced resistance [22]. Finally, the activity of chitosan as antitranspirant agent, via an ABA-mediated stomatal closure, may contribute to relieve the physiological stress due to water deficit in drought conditions [90]. Therefore, an approach based on chitosan application may really represent a sustainable, and possibly effective, strategy to control (virus) diseases in plant, by reducing the environmental impact of agrochemicals and their economic costs, possibly with higher crop yields and healthier plant products [91].

Acknowledgments We apologise to the colleagues whose excellent studies have not been cited for brevity.

MI is grateful to Dr Andrea Kuthanova and Prof Zdenek Opatrny (Department of Plant Physiology, Charles University in Prague, Czech Republic).

References

1. Agrios GN. Plant pathology. 5th ed. San Diego, CA: Elsevier Academic Press; 2006.
2. Stange C. Plant-virus interactions during the infective process. Cienc Investig Agrar. 2006;33(1):1–18.
3. Fraile A, García-Arenal F. The coevolution of plants and viruses: resistance and pathogenicity. Adv Virus Res. 2010;76:1–32.
4. Faoro F, Iriti M. Induction of resistance to virus diseases. Petria. 2009;19:149–72.
5. Iriti M, Faoro F. Constitutive and inducible immunity in plants. Petria. 2003;13:77–103.
6. Sanabria NM, Huang JC, Dubery IA. Self/nonself perception in plants in innate immunity and defence. Self Nonself. 2010;1(1):40–54.
7. Głowacki S, Macioszek VK, Kononowicz A. R proteins as fundamentals of plant innate immunity. Cell Mol Biol Lett. 2011;16:1–24.
8. Henry G, Thonart P, Ongena M. PAMPs; MAMPs; DAMPs and others: an update on the diversity of plant immunity elicitors. Biotechnol Agron Soc Environ. 2012;16:257–68.
9. Iriti M, Faoro F. Review of innate and specific immunity in plants and animals. Mycopathologia. 2007;164:57–64.
10. Kumar D, Klessig DF. The search for the salicylic acid receptor led to discovery of the SAR signal receptor. Plant Signal Behav. 2008;3:691–2.
11. Iriti M, Faoro F. Chitosan as a MAMP, searching for a PRR. Plant Signal Behav. 2009;4:66–8.
12. Chirkov SN. The antiviral activity of chitosan. Appl Biochem Microbiol. 2002;38:1–8.
13. Chen H-P, Xu LL. Isolation and characterization of a novel chitosan-binding protein from non-heading Chinese cabbage leaves. J Integr Plant Biol. 2005;47:452–6.
14. El Hadrami A, Adam LR, El Hadrami I, Daayf F. Chitosan in plant protection. Mar Drugs. 2010;8:968–87.
15. Yin H, Zhao X, Du Y. Oligochitosan: a plant disease vaccine—a review. Carbohydr Polym. 2010;82:1–8.
16. Falcón-Rodríguez AB, Wégria G, Cabrera J-C. Exploiting plant innate immunity to protect crops against biotic stress: chitosaccharides as natural and suitable candidates for this purpose. In: Bandani ER, editor. New perspective in plant protection. Rijeka: InTech; 2012.
17. Pochanavanich P, Suntornsuk W. Fungal chitosan production and its characterization. Lett Appl Microbiol. 2002;35:17–21.
18. Ylitalo R, Lehtinen S, Wuolijoki E, Ylitalo P, Lehtimäki T. Cholesterol-lowering properties and safety of chitosan. Arzneimittelforschung. 2002;52:1–7.
19. Kong M, Chen XG, Xing K, Park HJ. Antimicrobial properties of chitosan and mode of action: a state of art review. Int J Food Microbiol. 2010;144:51–63.
20. Henández-Lauzardo AN, Velázquez-del Valle MG, Guerra-Sánchez MG. Current status of action mode and effect of chitosan against phytopathogens fungi. Afr J Microbiol Res. 2011;5:4243–7.
21. Badawy MEI, Rabea EI. A biopolymer chitosan and its derivatives as promising antimicrobial agents against plant pathogens and their applications in crop protection. Int J Carbohydr Chem. 2011. doi:10.1155/2011/460381.
22. Iriti M, Castorina G, Vitalini S, Mignani I, Soave C, Fico G, Faoro F. Chitosan-induced ethylene-independent resistance does not reduce crop yield in bean plants. Biol Control. 2010;54:241–7.
23. Hadwiger LA, Beckman JM. Chitosan as a component of pea F. solani interactions. Plant Physiol. 1980;66:205–2011.
24. Pospieszny H, Atabekov JG. Effect of chitosan on the hypersensitive response reaction of bean to Alfalfa Mosaic Virus. Plant Sci. 1989;62:29–31.
25. Pospieszny H, Chirkov SN, Atabekov JG. Induction of antiviral resistance in plants by chitosan. Plant Sci. 1991;79:63–8.
26. Pospieszny H. Antiviroid activity of chitosan. Crop Prot. 1997;16:105–6.

27. Kulikov SN, Chirkov SN, Il'ina AV, Lopatin SA, Varlamov VP. Effect of molecular weight of chitosan on its antiviral activity in plants. Appl Biochem Microbiol. 2006;42:200–3.
28. Davydova VN, Nagorskaya VP, Gorbach VI, Kalitnik AA, Reunov AV, Solov'eva TF, Ermak IM. Chitosan antiviral activity: dependence on structure and depolymerization method. Appl Biochem Microbiol. 2011;47:103–8.
29. Pospieszny H, Struszczyk H, Chirkov SN, Atabekov JG. New application of chitosan in agriculture. In: Karnicki ZS, editor. Chitin world. Gdynia: Wirschaftsverlag NW; 1995.
30. Pospieszny H, Struszczyk H, Cajca M. Biological activity of *Aspergillus*-degraded chitosan. Chitin Enzymol. 1996;2:385–9.
31. Struszczyk MH, Pospieszny H, Schanzenbach D, Peter MG. Biodegradation of chitosan. In: Struszchyk H, editor. Progress on chemistry and application of chitin and its derivatives. Lodz, Poland: Polish Chitin Society; 1998.
32. Chirkov SN, Surguchova N, Gamzazade AI, Abdulabekov IM, Pospieszny HA. Comparative efficiency of chitosan derivatives as inhibitors of viral infection in plants. Doklady Rossi Akademii Nauk. 1998;360:271–3.
33. Chirkov SN, Il'ina AV, Surgucheca NA, Letunova EV, Varitsev YA, Tatarinova NY, Varlamov VP. Effect of chitosan on systemic viral infection and some defense responses in potato plants. Russ J Plant Physiol. 2001;48:774–9.
34. Rhoades J, Roller S. Antimicrobial actions of degraded and native chitosan against spoilage organisms in laboratory media and food. Appl Environ Microbiol. 2000;66:80–6.
35. Cabrera JC, Van Cutsem P. Preparation of chitooligosaccharides with degree of polymerization higher than 6 by acid or enzymatic degradation of chitosan. Biochem Eng J. 2005;25:165–72.
36. Chirkov SN, Surguchova N, Atabekov JG. Chitosan inhibits systemic infections caused by DNA-containing plant viruses. Arch Phytopathol Plant Protect. 1994;29:21–4.
37. Pospieszny H. Inhibition of tobacco mosaic virus (TMV) infection by chitosan. Phytopathol Polonica. 1995;22:69–74.
38. Surguchova NA, Varitsev YA, Chirkov SN. The inhibition of systemic viral infections in potato and tomato plants by chitosan treatment. J Russ Phytopathol Soc. 2000;1:59–62.
39. Lizama-Uc G, Estrada-Mota IA, Caamal-Chan MG, Souza-Perera R, Oropeza-Salìn C, Islas-Flores I, Zuñiga-Aguillar JJ. Chitosan activates a MAP-kinase pathway and modifies abundance of defence-related transcripts in calli of *Cocos nucifera* L. Physiol Mol Plant Pathol. 2007;70:130–41.
40. Baureithel K, Felix G, Boller T. Specific, high affinity binding of chitin fragments to tomato cells and membranes, competitive inhibition of binding by derivatives of chitin fragments and a Nod factor of Rhizobium. J Biol Chem. 1994;269:17931–8.
41. Shibuya N, Ebisu N, Kamada Y, Kaku H, Cohn J, Ito Y. Localization and binding characteristics of a high-affinity binding site for N-acetylchitooligosaccharide elicitor in the plasma membrane from suspension-cultured rice cells suggest a role as a receptor for the elicitor signal at the cell surface. Plant Cell Physiol. 1996;37:894–8.
42. Day RB, Okada M, Ito Y, Tsukada K, Zaghouani H, Shibuya N, Stacey G. Binding site for chitin oligosaccharides in the soybean plasma membrane. Plant Physiol. 2001;126:1162–73.
43. Okada M, Matsumura M, Ito Y, Shibuya N. High-affinity binding proteins for N-acetyl-chitooligosaccharide elicitor in the plasma membranes from wheat, barley and carrot cells: conserved presence and correlation with the responsiveness to the elicitor. Plant Cell Physiol. 2002;43:505–12.
44. Kaku H, Nishizawa Y, Ishii-Minami N, Akimoto-Tomiyama C, Dohmae N, Takio K, Minami E, Shibuya N. Plant cells recognize chitin fragments for defence signalling through a plasma membrane receptors. Proc Natl Acad Sci U S A. 2006;103:11086–91.
45. Hamel L-P, Beaudoin N. Chitooligosaccharide sensing and downstream signalling: contrasted outcomes in pathogenic and beneficial plant-microbe interactions. Planta. 2010;232:787–806.
46. Miya A, Albert P, Shinya T, Desaki Y, Ichimura K, Shirasu K, Narusaka Y, Kawakami N, Kaku H, Shibuya N. CERK1, a LysM receptor kinase, is essential for chitin elicitor signaling in Arabidopsis. Proc Natl Acad Sci U S A. 2007;104:19613–8.

47. Wan J, Zhang XC, Neece D, Ramonell KM, Clough S, Kim SY, Stacey MG, Stacey G. A LysM receptor-like kinase plays a critical role in chitin signalling and fungal resistance in Arabidopsis. Plant Cell. 2008;20:471–81.

48. Iizasa E, Mitsutomi M, Nagano Y. Direct binding of a plant LysM receptor-like kinase, LysM RLK1/CERK1, to chitin in vitro. J Biol Chem. 2010;285:2996–3004.

49. Petutschnig EK, Jones AM, Serazetdinova L, Lipka U, Lipka V. The lysin motif receptor-like kinase (LysM-RLK) CERK1 is a major chitin-binding protein in Arabidopsis thaliana and subject to chitin-induced phosphorylation. J Biol Chem. 2010;285:28902–11.

50. Zuppini A, Baldan B, Millioni R, Favaron F, Navazio L, Mariani P. Chitosan induces Ca^{2+} mediated programmed cell death in soybean cells. New Phytol. 2003;161:557–68.

51. Amborabé B-E, Bonmort J, Fleurat-Lessard P, Roblin G. Early events induced by chitosan on plant cells. J Exp Bot. 2008;59:2317–24.

52. Köhle H, Jeblick W, Poten F, Blaschek W, Kauss H. Chitosan-elicited callose synthesis in soybean cells as a Ca^{++}-dependent process. Plant Physiol. 1985;77:544–51.

53. Kauss H, Jeblick W, Domard A. The degree of polymerization and N-acetylation of chitosan determine its ability to elicit callose formation in suspension-cultured cells and protoplasts of Catharanthus roseus. Planta. 1989;178:385–92.

54. Iriti M, Sironi M, Gomarasca S, Casazza AP, Soave C, Faoro F. Cell death-mediated antiviral effect of chitosan in tobacco. Plant Physiol Biochem. 2006;4:893–900.

55. Faoro F, Maffi D, Cantù D, Iriti M. Chemical-induced resistance against powdery mildew in barley: the effects of chitosan and benzothiadiazole. Biocontrol. 2008;53:387–401.

56. Faoro F, Sant S, Iriti M, Maffi D, Appiano A. Chitosan-elicited resistance to plant viruses: a histochemical and cytochemical study. In: Muzzarelli RAA, editor. Chitin enzymology. Italy: Atec; 2001.

57. Faoro F, Iriti M. Cell death or not cell death: two different mechanisms for chitosan and BTH antiviral activity. IOBC/WPRS Bull. 2006;28:25–9.

58. Faoro F, Iriti M. Callose synthesis as a tool to screen chitosan efficacy in inducing plant resistance to pathogens. Caryologia. 2007;60:121–4.

59. Levy A, Guenoune-Gelbart D, Epel BL. β-1,3-Glucanases. Plasmodesmal gate keepers for intercellular communication. Plant Signal Behav. 2007;2:288–90.

60. Guenoune-Gelbart D, Elbaum M, Sagi G, Levy A, Epel BL. Tobacco mosaic virus (TMV) replicase and movement protein function synergistically in facilitating TMV spread by lateral diffusion in the plasmodesmal desmotubule of Nicotiana benthamiana. Mol Plant Microbe Interact. 2008;21:335–45.

61. Carrington JG, Kasschau KD, Mahajan SK, Schaad MC. Cell-to-cell and long-distance transport of viruses in plants. Plant Cell. 1996;8:1669–81.

62. Fiil BK, Petersen K, Petersen M, Mundy J. Gene regulation by MAP kinase cascades. Curr Opin Plant Biol. 2009;12:615–21.

63. Hu XY, Steven JN, Fang JY, Cai WM, Tang ZC. Mitogen activated protein kinases mediate the oxidative burst and saponin synthesis induced by chitosan in cell cultures of Panax ginseng. Sci China C. 2004;47:303–12.

64. Feng B, Chen Y, Zhao C, Zhao X, Bai X, Du Y. Isolation of a novel Ser/Thr protein kinase gene from oligochitosan-induced tobacco and its role in resistance against tobacco mosaic virus. Plant Physiol Biochem. 2006;44:596–603.

65. Yafei C, Yong Z, Xiaoming Z, Peng G, Hailong A, Yuguang D, Yingrong H, Hui L, Yuhong Z. Functions of oligochitosan induced protein kinase in tobacco mosaic virus resistance and pathogenesis related proteins in tobacco. Plant Physiol Biochem. 2009;47:724–31.

66. Mandal S, Mitra A. Reinforcement of cell wall in roots of Lycopersicon esculentum through induction of phenolic compounds and lignin by elicitors. Physiol Mol Plant Pathol. 2007;71:201–9.

67. Buonaurio R, Iriti M, Romanazzi G. Induced resistance to plant diseases caused by oomycetes and fungi. Petria. 2010;19:125–49.

68. Zaninotto F, La Camera S, Polverari A, Delledonne M. Cross talk between reactive nitrogen and oxygen species during the hypersensitive disease resistance response. Plant Physiol. 2006;141:379–83.
69. Zhao XM, She XP, Yu W, Liang XM, Du YG. Effects of oligochitosans on tobacco cells and role of endogenous nitric oxide burst in the resistance of tobacco to TMV. J Plant Pathol. 2007;89:55–65.
70. Li Y, Yin H, Qing W, Zhao X, Du Y, Li F. Oligochitosan induced *Brassica napus* L. production of NO and H_2O_2 and their physiological function. Carbohydr Polym. 2009;75:612–7.
71. Srivastava N, Gonugunta VK, Puli MR, Raghavendra AS. Nitric oxide production occurs downstream of reactive oxygen species in guard cells during stomatal closure induced by chitosan in abaxial epidermis of *Pisum sativum*. Planta. 2009;229:757–65.
72. Manjunatha G, Roopa KS, Prashanth GN, Shetty HS. Chitosan enhances disease resistance in pearl millet against downy mildew caused by Sclerospora graminicola and defence-related enzyme activation. Pest Manag Sci. 2008;64:1250–7.
73. Lin W, Hu X, Zhang W, Rogers WJ, Cai W. Hydrogen peroxide mediates defence responses induced by chitosans of different molecular weights in rice. J Plant Physiol. 2005;162:937–44.
74. Zhang H, Wang W, Yin H, Zhao X, Du Y. Oligochitosan induces programmed cell death in tobacco suspension cells. Carbohydr Polym. 2012;87:2270–8.
75. Wang W, Li S, Zhao X, Du Y, Lin B. Oligochitosan induces cell death and hydrogen peroxide accumulation in tobacco suspension cells. Pestic Biochem Physiol. 2008;90:106–13.
76. Vasil'ev LA, Dzyubinskaya EV, Zinovkin RA, Kiselevsky DB, Lobysheva NV, Samuilov VD. Chitosan-induced programmed cell death in plants. Biochemistry (Moscow). 2009;74:1035–43.
77. Malerba M, Crosti P, Cerana R. Defense/stress responses activated by chitosan in sycamore cultured cells. Protoplasma. 2012;249:89–98.
78. Doares SH, Syrovets T, Weiler EW, Ryan CA. Oligogalacturonides and chitosan activate plant defence genes through the octadecanoid pathway. Proc Natl Acad Sci U S A. 1995;92:4095–8.
79. Rakwal R, Tamogami S, Agrawal GK, Iwahashi H. Octadecanoid signalling component "burst" in rice (*Oryza sativa* L.) seedling leaves upon wounding by cut and treatment with fungal elicitor chitosan. Biochem Biophys Res Commun. 2002;295:1041–5.
80. Obara N, Hasegawa M, Kodama O. Induced volatiles in elicitor-treated and rice blast fungus-inoculated rice leaves. Biosci Biotechnol Biochem. 2002;66:2549–59.
81. Ogawa D, Nakajima N, Seo S, Mitsuhara I, Kamada H, Ohashi Y. The phenylalanine pathway is the main route of salicylic acid biosynthesis in Tobacco mosaic virus-infected tobacco leaves. Plant Biotechnol. 2006;23:395–8.
82. Iriti M, Faoro F. Abscisic is involved in chitosan-induced resistance to tobacco necrosis virus (TNV). Plant Physiol Biochem. 2008;2008(46):1106–11.
83. Yin H, Bai XF, Zhao XM, Du YG. Molecular cloning and characterization of a *Brassica napus* L. MAP kinase involved in oligochitosan-induced defence signalling. Plant Mol Biol Rep. 2010;28:292–301.
84. Walker-Simmons M, Jin D, West CA, Hadwiger L, Ryan CA. Comparison of proteinase inhibitor-inducing activities and phytoalexin elicitor activities of a pure fungal endopolygalac-turonase, pectic fragments and chitosans. Plant Physiol. 1984;76:833–6.
85. Park CJ, Kim KJ, Shin R, Park JM, Shin YC, Paek KH. Pathogenesis-related protein 10 isolated from hot pepper functions as a ribonuclease in an antiviral pathway. Plant J. 2004;37:186–98.
86. Uthairatanakij A, Teixeira da Silva JA, Obsuwan K. Chitosan for improving orchid production and quality. Orchid Sci Biotechnol. 2007;1:1–5.
87. Iriti M, Vitalini S, Di Tommaso G, D'Amico S, Borgo M, Faoro F. A new chitosan formulation induces grapevine resistance against powdery mildew and improves grape quality traits. Aust J Grape Wine Res. 2011;17:263–9.

88. Vitalini S, Gardana C, Zanzotto A, Fico G, Faoro F, Simonetti P, Iriti M. From vineyard to glass: agrochemicals enhance the melatonin content, total polyphenols and antiradical activity of red wines. J Pineal Res. 2011;51:278–85.
89. Khan MR, Doohan FM. Comparison of the efficacy of chitosan with that of a fluorescent pseudomonad for the control of Fusarium head blight disease of cereals and associated mycotoxin contamination of grain. Biol Control. 2009;48:48–54.
90. Iriti M, Picchi V, Rossoni M, Gomarasca S, Ludwig N, Gargano M, Faoro F. Chitosan antitranspirant activity is due to abscisic acid-dependent stomatal closure. Exp Environ Bot. 2009;66:493–500.
91. Mejía-Teniente L, Torres-Pacheco I, González-Chavira MM, Ocampo-Velazquez RV, Herrera-Ruiz G, Chapa-Oliver AM, Guevara-González RG. Use of elicitors as an approach for sustainable agriculture. Afr J Biotechnol. 2010;9:9155–62.

Chapter 7
The Application of Oligosaccharides as Plant Vaccines

Xiaoming Zhao, Wenxia Wang, Yuguang Du, and Heng Yin

Abstract Plant diseases and insect pests constitute one of the main factors of restricted agriculture development. To resist the damage caused by pathogens and pests, plants have evolved very sophisticated mechanisms of immunity to combat them using as little reserved or generated energy as possible. The immune mechanisms of plants can be activated by outside elicitors, and may be applied to control plant disease. An oligosaccharide is a type of elicitor, also called plant vaccine, that has a beneficial effect on agricultural production. The author has summarized the effects of oligochitosan on crop disease control, preventing chilling injury, and promoting plant growth in field experiments. Oligochitosan is multifunctional plant immune vaccine, It can effectively control crop diseases at 50 parts per million (ppm), effectively prevent chilling injury caused by the late spring coldness at 75 ppm, and can promote the growth of crops. These results show that there are bright prospects for the application and study of oligosaccharins.

Keywords Oligosaccharides • Application • Plant vaccines • Field experiment

Introduction

Plant diseases and pests have a great impact on agricultural development, accounting for 20–30 % of total losses. Over-reliance on chemical pesticides to control pest and plant disease has led to pathogens and insects becoming resistant to

X. Zhao, Ph.D. • W. Wang, Ph.D. • H. Yin, Ph.D. (✉)
Biotechnology Department, Dalian Institute of Chemical Physics,
Chinese Academy of Sciences, Dalian 116023, China
e-mail: zhaoxm@dicp.ac.cn; yinheng@dicp.ac.cn

Y. Du, Ph.D.
Biotechnology Department, Dalian Institute of Chemical Physics,
Chinese Academy of Sciences, Dalian 116023, China

Institute of Process Engineering, Chinese Academy of Sciences, Beijing, China
e-mail: ygdu@ipe.ac.cn

© Springer Science+Business Media New York 2016
H. Yin, Y. Du (eds.), *Research Progress in Oligosaccharins*,
DOI 10.1007/978-1-4939-3518-5_7

83

conventional pesticides, stimulating the new physiological races of pathogen that are emerging; thus, plant disease and insect damage have become more rampant. Therefore, the amount of chemical pesticides used is increasing. Excessive use of chemical pesticides has caused serious pollution of agriculture products and the environment, and has affected people's physical and mental health. At present, for plant virus disease, soil-borne disease, and plant nematodes, there are still no effective prevention and control of reagents. There is an urgent need to study a new approach to protecting and controlling diseases.

Faced with viruses, bacteria, fungi, parasitic plants, nematodes, and other diseases, plants have evolved different defense responses, such as the production of anti-toxins, protease inhibitors, hydrolytic enzymes (chitinase, β-1.3 glucanase, etc.), glycoprotein on the cell wall, and lignification [1–3]. These defense responses can be induced by elicitors. In 2008, Qiu et al. [4] put forward the concept of a plant vaccine. Yin et al. [5] reported that oligosaccharides constitute one effective plant vaccine.

Research into oligosaccharides began in the 1960s. In 1976, Ayers et al. [6] found that phytoalexin can be induced by the low molecular weight of oligosaccharide from the fungal cell wall. Albersheim and Darvill [7] first proposed that oligosaccharins could not only regulate plant growth, development, and reproduction, but also increase the plant's responses, which could stimulate the immune system of the plants. Oligosaccharins may trigger defense responses and regulate plant growth, followed by the production of active substances to inhibit disease formation. In recent years, studies have found that chitosan oligosaccharides (COS) could not only induce plant disease resistance, but also stimulate the resistance to abiotic stresses, including drought and cold environments. Furthermore, the growth of plants can be triggered by COS, which shows the prospect of a wide application in agricultural production.

The Application of Oligosaccharides to Crops to Protect and Control Diseases

Oligosaccharides for Protecting and Controlling Fungal Diseases

Rice Diseases

Widely distributed rice blast and sheath blight pose great threats to rice production, causing an annual loss of billions of kilograms, even taking into account a 50 % drop in areas of high incidence. Ning et al. [8] sprayed 5 μg/mL of COS to rice H7R (resistance) as a positive control and H7S (susceptible) respectively, followed by inoculation of the spores of rice blast (*Magnaporthe grisea* 01-19B, small species) 8 h later. It was shown that COS clearly improved rice resistance to rice blast disease as much as 50 %, as the lesions declined and the infection rate slowed down. Meanwhile, COS induced hypersensitive responses (HR) along with H_2O_2 production in rice.

Bai et al. [9] observed the effect of COS on rice sheath blight. The results showed that there was no obvious inhibitory effect on the mycelium and sclerotia of rice sheath blight diseases. However, having sprayed COS on adult rice leaves, the incidence of the disease index in rice was significantly lower than the control, where different concentrations led to different effects, the strongest resistance effect being 65.56 % at a concentration of 50 µg/mL. The activities of some enzymes, such as rice plant peroxidase (POD), polyphenol oxidase (PPO), and β-1,3-glucanase improved to varying degrees after the application of COS.

Hu et al. [10] investigated whether COS with different molecular weights (MWs) have an impact on the species of Lemont, Teqing, and Qimiaoxiang against sheath blight (*Rhizoctonia solani*). They concluded that an MW of 1500 possessed the best induction effect on chitinase, whereas MW of 500 showed the worst.

Wheat Diseases

Zhang et al. [11] reported that COS provided good protection against and control of wheat sheath blight, ranging from 88.40 to 90.60 % according to the results of indoor and field experiments. Liu et al. [12] reached a conclusion that the pretreatment of wheat seeds with COS could remove the inhibitory effect of deoxynivalenol (DON) produced by *Fusarium graminearum* from etiolated wheat seedlings and embryo cells, indicating that oligosaccharides improved plant resistance to pathogen toxins.

Cucumber Diseases

Chitosan and chitin oligomers were used by Ben-Shalom et al. [13] to protect and control cucumber diseases. As shown in Fig. 7.1, the former was more effective than the latter. After the inoculation of chitosan 24 h later, the disease index of chitosan

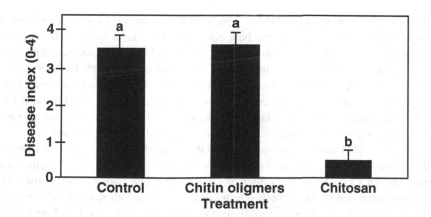

Fig. 7.1 The effect of chitosan and chitin oligomers on *Botrytis cinerea*

was 0.45, with a protective effect of 87.14%, whereas the control disease index was 3.5, with no obvious differences compared with chitin oligomers.

After a 24-h pretreatment of water, chitin oligomer, and chitosan, spores of *B. cinerea* were inoculated later. The disease index was investigated 5 days later at 22 °C.

The inhibitory effect of chitosan against spore germination and the length of the germ tube of *B. cinerea* have been investigated. The results showed that chitin has no effect on the fungus. However, the inhibitory rate of a low concentration (20–30 µg/mL) of chitosan was as high as 50%, whereas that of chitosan with a concentration of 50 µg/mL almost controlled the growth of *B. cinerea*. Twenty-four hours later, the length of spore under chitosan at 10 µg/mL was 2 µm, whereas that with water pretreatment was 15 µm on average.

With transmission electron microscopy, Ma et al. [14] researched the inhibitory effect of oligosaccharides against powdery mildew *Sphaerotheca fuliginea* in the leaves of cucumber *Cucumis sativus*. The ultrastructural observations showed that bacteria growth was obviously inhibited after the pretreatment. The condensation of mycelium cytoplasm was observed, along with organelle disintegration and the collapse of cells and tissues. The electron density of the protoplasm was deeper inside and the inhaler became deformed, the wall suction thickened, combined with organelle disintegration, ultimately followed haustorium necrosis. On the basis of the previous work, the authors concluded that the chitosan induced defense reactions against *S. fuliginea* in cucumber [15]. They found that the pretreatment of oligosaccharides on the leaves could lead to the alleviation of diseases, with sparse and smaller lesions, the delay of sporulation, and the extension of the latent period. The best treatment period was 5~7 days before inoculation, and the persistence of induced resistance could last 10 d or more. After the pretreatment of elicitors on the first true leaves, the upper leaves displayed apparent resistance, indicating that oligosaccharides have the ability to induce systemic acquired resistance against powdery mildew in cucumber seedlings.

Pepper Diseases

Xiao et al. [16] reported that COS was able to induce resistance to powdery mildew in pepper at 25–125 µg/mL, with the best effect of 80% after a 1-day pretreatment at 50 µg/mL. Meanwhile, they also found that COS induced systemic acquired resistance against powdery mildew in pepper [17]. After pretreatment, the activities of polyphenol oxidase (PPO), peroxidase (POD), and phenylalanine ammonia lyase (PAL) were distinctly increased compared with the controls, in which PPO and PAL peaks occurred early (after inoculation). It was shown that the activities of the enzymes were in accordance with systemic acquired resistance after COS pretreatment.

Xu et al. [18] found that spraying COS in the field has a great impact on *Phytophthora capsici*, the effect of which was as high as 73.2% at a 40µg/mL concentration of COS. From the effect of COS on the mycelial growth in the PDA medium according to the growth rate method, it was shown that amino oligosaccharins were able to inhibit mycelial growth at EC50 to 100 µg/mL, whereas COS inhibits sporangia formation and stationary spore germination of *P. capsici* with effective EC50 of 0.64 µg/mL and 41.84 µg/mL respectively.

Oilseed Rape (*Brassica napus H*) Diseases

Yin et al. [19] conducted the experiments using COS on *Sclerotinia sclerotiorum* and found that the effect of COS was time-dependent. Before inoculation of the fungus, a 3-day pretreatment of 50 µg/mL COS led to the best control effect of as high as 72.1 %. However, it was demonstrated that systemic acquired resistance of oilseed rape was triggered by the treatment of COS instead of the direct effect of COS on the *S. sclerotiorum*. Semi-quantitative RT-PCR detected that one of the most important resistance genes, BnPDF 1.2, could be up-regulated, along with an increase in one of the most important enzymes, lipoxygenase (LOX), on the jasmonic acid (JA) pathway. All indicated that COS was likely to induce oilseed rape against *S. sclerotiorum* via the JA/ET-mediated pathway.

Tomato Diseases

In 2001, scientists from the Shaanxi Institute of Plant Protection demonstrated that COS played a vital role in the efficacy of tomato early blight. Compared with the effect of 68.79 % using the control mancozeb (70 %) after the first 7 days of treatment, the effects of COS were 67.25 % at 20 µg/mL, 71.29 % at 25 µg/mL, and 79.40 % at 40 µg/mL respectively. Seven days after the second treatment, compared with 70.63 % using the control, the effects of COS were 67.99 % at 20 µg/mL, 77.50 % at 25 µg/mL, and 80.15 % at 40 µg/mL. Seven days after the second treatment, the effects of COS were 70.82 % at 20 µg/mL, 77.37 % at 25 µg/mL, and 84.47 % at 40 µg/mL, in comparison with 73.27 % using the control (unpublished).

Fusarium wilt in Watermelon

Xu et al. [20] reached a conclusion that COS has a great impact on the protection of watermelon from *F. wilt* and on promotion of the growth of watermelon seedlings. The effects of protection and control were 63.98 % indoors and 71.82 % in the field. Furthermore, plant height, root length, and fresh weight increased significantly. From the field results, the authors found that COS treatment was 52.60 % more effective than the control.

Control of Crop Virus Diseases by Oligosaccharide Elicitors

Control of Virus Diseases in Legumes

Oligosaccharides induce significant plant resistance not only against plant fungal diseases, but also against plant viral diseases. Pospieszny et al. [21] first reported that chitosan inhibited infection of beans with the alfalfa mosaic virus (AMV) in 1989. Pospieszny et al. [22] reported that chitosan induced resistance to alfalfa

Table 7.1 Induced resistance of chitosan to different viruses in plants

Plant	Virus	Effects	Inhibition
Phaseolus vulgaris	AMV-L	Decrease the number of necroses	++++
Phaseolus vulgaris	TNV		++++
Chenopodium quinoa	TNV		++
Chenopodium quinoa	CMV		+++
Nicotiana tabacum	TMV		+++
Var. Samsun			
Nicotiana tabacum	TMV		++
Xanthi nc			
Nicotiana glutinosa	TMV		++
Nicotiana paniculata	PSV		++
Phaseolus vulgaris	AMV-S	Decrease the number of infected	++++
Phaseolus vulgaris	PSV	plants	++++
Pisum sativum	AMV		+++
Pisum sativum	PSV		+++
Lycopersicum esculentum	PVX		+++

All plants were inoculated after 0.1 % chitosan treatment for 1 day
Plants systemically infected were examined by ELISA 10–12 days after inoculation
AMV alfalfa mosaic virus, *TNV* tobacco necrosis virus, *CMV* cucumber mosaic virus, *TMV* tobacco mosaic virus, *PSV* panicum mosaic satellite virus, *PVX* potato virus X

mosaic virus (AMV), tobacco necrosis virus (TNV), tobacco mosaic virus (TMV), peanut stunt virus (PSV), cucumber mosaic virus (CMV), and potato virus X (PVX) in various plants (Table 7.1). These findings suggested that chitosan inhibiting viral infection might be associated with the relationship of the virus to the plant, chitosan concentration, and methods of chitosan treatment. It is obvious to note that the treatment of bean leaves with chitosan decreased the number of local necroses caused by infection with AMV and TMV. Inhibition of TMV infection is dependent on chitosan concentration, which is negative correlation. Furthermore, TMV multiplication was inhibited completely after inoculation 6–8 h before 0.01 % chitosan treatment in tobacco protoplast according to ELISA detection. It is illustrated that chitosan not only inhibits virus infection, but also inhibits virus multiplication in cells.

Control of Tobacco Virus Diseases

Shang et al. [23] reported that oligochitosan induced resistance to TMV in tobacco. It was shown that oligochitosan at a concentration of 50 µg/mL inhibited TMV infection to 84.73 % after pretreatment 24 h before TMV inoculation. Oligochitosan pretreatment decreased starch stains in half of tobacco leaves by KI-I staining. Furthermore, the chlorophyll content was 8.67 µg/g after treatment with 50 µg/mL oligochitosan, which was higher than after TMV treatment and after virus A treatment. Shang et al. [24] also reported that oligochitosan inhibited TMV multiplication in *Nicotiana tabacum*. It was shown that pretreatment with 50 µg/mL oligochitosan induced systemic resistance in tobacco. The viral disease was delayed for 4–7 days, and the average severity decreased to 82.9 %. The TMV content was reduced in tobacco plants after oligochitosan treatment according to ELISA-DSM detection. For example, the OD value was 0.400 in inoculated leaves in the oligochitosan group 10 days after inoculation, which was 23 % of that of the water treatment group. The OD value was 0.190 in new leaves of the oligochitosan group, which was 38.7 % of that of the water treatment group. All these results showed that oligochitosan inhibited TMV multiplication in *Nicotiana tabacum*. The effect of oligochitosan induction on the long-distance movement of TMV in tobacco was reported by Shang et al. [25]. The results indicated that a long-distance downward movement and upward movement of TMV were delayed or inhibited and the downward movement became more inhibited. In 2007, Shang et al. [26] reported that TMV-CP gene expression decreased significantly after oligochitosan administration in tobacco. It was reported by Shang et al. [27] that oligochitosan could inactivate TMV particles *in vitro*, which showed that TMV pathogenicity was decreased by combining with oligochitosan. It revealed that treatment with 300 µg/mL oligochitosan directly broke 80 % of TMV particles into tiny pieces of 50–150 nm according to JEM-1230 transmission electron microscope observation, which suggested that oligochitosan destroyed TMV particles with strength. Shang et al. [28] reported that 50 µg/mL oligochitosan induced activities of the defense-related enzymes SOD, POD, PAL, and PR-1a expression in tobacco. Zhao et al. [29] reported control of tobacco virus disease by oligochitosan in the field, which demonstrated the effective antiviral activity of oligochitosan, and the control effect was 77.9 %.

Control of Tomato and Pepper Virus Diseases

In 2000, the Institute of Pesticide Verification in Liaoning found that tomato virus disease could be inhibited by oligochitosan treatment at concentrations of 40 µg/mL, 50 µg/mL, and 60 µg/mL, and the control effects were 51.6 %, 64.7 %, and 68.2 % respectively (unpublished). Zhao et al. [29] reported control of tomato virus disease by oligochitosan in the field, demonstrating control effects of 74.45 %, 70.8 %, 63.05 % at concentrations of oligochitosan of 40 µg/mL, 50 µg/mL, and 60 µg/mL, and the growth rates were 16.7 %, 15.0 %, and 10 % respectively.

The control effect of pepper virus disease was good after oligochitosan treatment. In 2001, the Institute of Plant Protection in Hainan found that oligochitosan could inhibit pepper virus disease, and the control effect was 88.69 % with 50 µg/mL oligochitosan treatment. Zhao et al. [29] reported control of pepper virus disease by oligochitosan in field, with control effects of 56.9 %, 69.8 %, 77.0 % at oligochitosan concentrations of 40 µg/mL, 50 µg/mL, and 60 µg/mL, and the growth rates were significant: 3.9 %, 10.1 %, and 18.2 % respectively.

Control of Crop Bacterial Diseases by Oligosaccharide Elicitors

There are fewer reports on oligosaccharide elicitors inducing plant resistance against bacterial diseases, but this has attracted attention. Some experiments on the inhibition of bacterial growth by chitosan have been completed, as reported by Rabea et al. (Table 7.2) [30]. It has been noted that the inhibitory effect of chitosan was different concentration depending on the type of bacteria. For example, chitosan significantly inhibited *Corynebacterium michiganense*, the lowest concentration being 10 parts per million (ppm), whereas the lowest effective concentration needed to inhibit *Bacillus cereus* growth was 1,000 ppm.

In 2000, the Institute of Plant Protection in Hainan found that oligochitosan could inhibit soft rot in Chinese cabbage, and the control effect was 85 % with oligochitosan treatment at a concentration of 60 μg/mL, whereas the control effect was 74.99 % with 72 % streptomycin sulfate at a dilution of 3,000 as a control. In 2001, it was shown that soft rot in Chinese cabbage was inhibited by oligochitosan in the field by the Institute of Vegetables and Flowers, Chinese Academy of Agricultural Sciences. The control effects were 66.90 %, 62.26 %, 47.60 % at concentrations of 60 μg/mL, 50 μg/mL, and 40 μg/mL oligochitosan. The plants were sprayed five times continuously before disease occurrence. In 2001, the Institute of Pesticide Verification in Liaoning reported that the control effect on soft rot in Chinese cabbage was 78.62 after 60 μg/mL oligochitosan treatment, and the growth rate was 16.67 %.

The Application of Oligosaccharide Elicitors to Activate Plant Cold Resistance

Crops could be injured by coldness to varying degrees. Some serious conditions will result in crop failure. Oligosaccharide elicitors can activate plants' resistance not only to disease but also to cold. Kuang et al. [31] treated the seeds of

Table 7.2 The minimum inhibitory concentrations (MIC) of chitosan

Bacteria	MIC (ppm)
Agrobacterium tumefaciens	100
Bacillus cereus	1,000
Corynebacterium michiganense	10
Erwinia spp.	500
Erwinia carotovora subsp.	200
Escherichia coli	20
Klebsiella pneumoniae	700
Micrococcus luteus	20
Pseudomonas fluorescens	500
Staphylococcus aureus	20
Xanthomonas campestris	500

rice (*Oryza sativa L.*) with 25, 50, 100, and 200 mg/L oligochitosan solutions and the effects of oligochitosan on the resistance to cold of rice seedlings were studied. It was found that although each treatment increased cold resistance, 100 mg/L oligochitosan treatment was much more effective. Compared with control group, 100 mg/L oligochitosan treatment increased the survival percentage of seedlings under cold stress by 15.54 %; electrolyte leakage and malondialdehyde (MDA) content decreased by 12.88%and 6.98 %; the activities of superoxide dismutase (SOD), catalase (CAT), and peroxidase (POD) increased by 37.94 %, 22.75%, and 21.35 % respectively; the content of soluble sugar, soluble protein, and praline increased by 15.82 %, 12.92 %, and 24.47 % respectively; and the content of chlorophyll and rooting ability increased 23.53% and 28.90 %. Kuang et al. [31] treated the seedlings of eggplants (*Solanurn melongena L.*) at the four-leaf stage with 1/1,500, 1/1,000, 1/500, and 1/100(w/v) of oligochitosan for 2 days and then exposed them to a low temperature of 5 °C for 3 days to study the effect of oligochitosan on chilling tolerance. The chilling tolerance of eggplant seedlings could be increased with the application of oligochitosan treatment. Compared with the controls, the activities of three protective enzymes, superoxide dismutase (SOD), peroxidase (POD), and catalase (CAT), increased, whereas the MDA content decreased in the leaves of treated plants. On the other hand, oligochitosan could effectively promote the increase in the content of proline and soluble sugar in the leaves of eggplant seedlings during low-temperature stress. A concentration of oligochitosan of 1/1,000 (w/v) produced the most effective result.

Crops produce small, abnormal fruit owing to freezing injury caused by coldness in the late spring. Some crops even stop producing completely. Coldness can decrease the fruit setting ratio of the crisp pear, which usually blooms in the late spring. Cold temperature was a very serious problem in 2007 in Shaanxi Province at Pucheng County. It caused failure of the crisp pear crop. DICP studied the effect of chito-oligosaccharides on plant cold resistance in Shaanxi Province at Pucheng County. The result shows that before the cold period in late spring, the crisp pear trees treated with a 75 mg/L concentration of oligochitosan expressed chilling tolerance activity, with a high fruit setting ratio. The result measured in 2009 shows that the trees treated with oligochitosan had a 9.9 times higher fruit setting ratio than those that had not been treated (Table 7.3). Besides the higher fruit setting ratio, the young fruits of trees treated with oligochitosan had grown faster (Table 7.4). In the research reported on 25 April 2009, the young fruits treated with oligochitosan had grown 1.42 cm in diameter. If young fruit encounter cold weather, treating with oligochitosan can protect the fruits from serious cold injury. In 2010, fruit encountered cold weather from 2 to 6 April. The young fruit treated with oligochitosan showed a strong cold tolerance. The cold injury ratio of fruit from trees that had been treated is much lower than that of fruit from untreated trees. Surveys showed that the fruit cold injury ratio of trees treated with oligochitosan was 20.5 %, whereas the ratio of trees that were untreated was 80.25 %. The area of cold injury was 1–5 % in the fruit, whereas in the untreated trees it was 10–30 %.

Table 7.3 The effect of chitosan on crisp pear and coldness

Treatment	Spraying time	Florescence	Average diameter of young fruits (mm)	Diameter adding (mm)
Chitosan (75 µg/mL)	March 26th	March 27th to April 12th	15.22	1.42
Control			13.80	–

Table 7.4 The effect of chitosan on crisp pear and fruit setting

Treatment	Number of flower buds	Number of fruit setting	Fruit setting ratio (%)	Increase in fruit setting (times)
Chitosan (75 µg/mL)	426	124	29.11	9.90
Control	450	12	2.67	–

The Application of Oligosaccharide to Promote Plant Growth

Besides the application of activating plant disease resistance, oligosaccharides have been applied to promote plant growth. Zhang et al. [32] proved that the chitoligmer can promote cucumber growth to a large extent in agriculture. The result shows that plants treated with chitoligmer have enhanced disease resistance, and the harvest time is 3–5 days earlier than for untreated plants. Guo et al. [33] used different concentrations of oligochitosan to treat tobacco seedlings. The results indicated that a concentration of 0.01 mg/L oligochitosan promoted the growth of tobacco seedlings, and that the heights of seedlings, the areas of functional leaves, the chlorophyll content, the net rate of photosynthesis (P_n), stomatal conductance (G_s), intercellular CO_2 concentration (C_i), and the transpiration rate (T_r)of tobacco seedlings were all increased, whereas stomatal limitation (L_s)was reduced. A high concentration of oligochitosan (100 mg/L) restricted the growth of tobacco seedlings. The optimal concentration of oligochitosan for tobacco seedlings was 0.01 mg/L. The effects of two oligochitosan treatments were better than those of a single treatment. Guo et al. [34] used different concentrations of oligochitosan on cucumber seeds and seedlings, and the results indicated that low concentrations of oligochitosan promoted the germination of cucumber, the optimal concentration being 0.1 mg/L. The same as for tobacco, a low concentration of oligochitosan also promoted the growth of cucumber seedlings, and the heights of seedlings, the areas of functional leaves, and the length of roots all increased significantly ($P<0.05$) compared with CK, the chlorophyll content, P_n, G_i, C_i, and T_r of cucumber seedlings were increased significantly ($P<0.05$), whereas L_s was reduced significantly ($P<0.05$). A high concentration of oligochitosan (100 mg/L) restricted the growth of cucumber seedlings too. The optimal concentration of oligochitosan for cucumber seedlings was 0.1 mg/L. The effect of using two oligochitosan treatments was more effective than a one-time treatment (Table 7.5) [33].

Table 7.5 Effects of times and concentrations of oligochitosan on the growth of tobacco

Chitosan concentration (mg/L)	Stem length (cm)		Stem diameter (cm)		Areas of function leaves (cm²)		Length of root (cm)	
	Treat once	Treat twice	Treat once	Treat twice	Treat once	Treat twice	Treat once	Treat twice
CK	10.6b	10.2b	0.338a	0.341a	36.12a	38.75a	12.6a	11.9a
0.001	10.9b	11.1b	0.329a	0.322a	38.19a	37.63a	12.1a	12.5a
0.01	15.3d	18.5e	0.341a	0.338a	47.51b	50.03b	14.6b	15.9b
1	13.2c	16.1d	0.3218a	0.326a	41.93a	47.77b	14.1b	15.3b
100	8.9a	8.3a	0.332a	0.334a	37.16a	35.28a	11.9a	11.5a

$P < 0.05$. Values in the same column with different letters are significantly different ($p < 0.05$). The letter 'a' denotes the lowest value.

Li et al. [35] used oligochitosan prepared from enzymatic hydrolysis of chitosan on the parameters of photosynthesis of *Brassica napus L.* leaves. To test the these parameters, *B. napus L.* seedlings with or without oligochitosan treatment were determined using PP-Systems CIRAS-2 portable photosynthetic apparatus in the trial. The result indicated that the P_n, G_s, and G_i increased significantly in *B. napus L.* leaves treated with 5 or with 50 mg/L oligochitosan ($P < 0.05$). It indicated that oligochitosan treatments are beneficial in increasing the parameters of photosynthesis of *B. napus L.* Oligochitosan treatment at a concentration of 5 mg/L was more effective than oligochitosan treatment of 50 mg/L. The P_n, G_s, and C_i decreased with the co-treatment of 5 mg/L oligochitosan and L-NAME or 5 mg/L oligochitosan and Na_2WO_4 together. The P_n, G_s, and C_i increased in *B. napus L.* leaves treated with NO or ABA. This indicated that oligochitosan treatment increased the parameters of photosynthesis of *B. napus L.* seedlings via the NO or the ABA pathway.

Prospects for the Application Oligosaccharide-Based Biopesticide

Plant activator can activate the disease defense system of plant, which has been studied for many years. However, the application of research achievements to agriculture is just beginning. With the development of research work, oligosaccharide biopesticides will play an important role in agriculture in the future.

The disease defense system induced by plant activators has a broad spectrum of antiviral activity, is of longer duration, and is better controlled in space and time. The most important thing is that nearly all the elicitors do not pollute the environment. All of these advantages presage the future extensive use of oligosaccharide biopesticides. From the biology elicitors to the chemistry elicitors, to the signal transduction caused by plant activators, methods of increasing the effect of the activators is also an important aspect for study. With the development of this aspect, less pesticide will be used, to avoid agricultural and food pollution. Large-scale application of oligosaccharide-based biopesticides in agriculture as great potential.

Oligosaccharins, a type of signal molecule, have only recently been discovered, but they are also the most in-depth studied molecules. Their functions are to regulate plant growth, propagation, and disease defense. Both the immune system and the defense reaction can be activated by oligosaccharins. Some substances with disease resistance accumulated. Because the oligosaccharides with different origins act on different pathogens, a series of oligosaccharide biopesticides can be developed to deal with the genetic variability of pathogens, which is difficult to solve using genetic breeding. In addition to the advantage mentioned above, oligosaccharide biopesticides have wide-reaching material sources, low production costs, and is effective as a non-toxic and pollution-free drug. Therefore, both oligosaccharide biopesticide-producing enterprises and agricultural production will unquestionably gain great economic benefit. The application of oligosaccharins to disease prevention in marine breeding and pasturage is a very promising prospect. With the development of glycobiology, there will be a bright future for the application and study of oligosaccharins.

References

1. Tuzun S, Bent E. Multigenic and induced systemic resistance in plants. Berlin: Springer Science+Business Media Inc; 2006.
2. Kuć J. Development and future direction of induced systemic resistance in plants. Crop Prot. 2000;19:859–61.
3. Sticher L, Much-Mani B, Métraux JP. Systemical acquired resistance. Annu Rev Phytopathol. 1997;35:235–70.
4. Qiu D, et al. Plant immunity and plant vaccine: research and practice. Beijing: Science Press; 2008.
5. Yin H, Zhao X, Du Y. Oligochitosan: a plant diseases vaccine – a review. Carbohydr Polym. 2010;82(1):1–8.
6. Ayers AR, Ebel J, Valent B, Albersheim PA. Host-pathogen interactions X. Fractionation and biological activity of elicitor isolated from the mycelial walls of Phytophthora megasperma var. sojae. Plant Physiol. 1976;57:760–5.
7. Albersheim P, Darvill AG. Oligosaccharins. Sci Am. 1985;253:58–64.
8. Ning W, Chen F, Mao B, Li Q, Liu Z, Gao Z, He Z. N-acetylchitooligosaccharides elicit rice defence responses including hypersensitive response-like cell death, oxidative burst and defence gene expression. Physiol Mol Plant Pathol. 2004;64:263–71.
9. Bai C, Jiang X, Ding H, Zuo X. Resistance induced by chitosan oligosaccharide to rice sheath blight. Guizhou Agric Sci. 2010;38(8):103–6.
10. Hu J, Chen Y, Chen Z. Chitinase induction with chitosan oligosaccharides on several rice varieties with different resistant ability to sheath blight (Rhizoctonal Solani K.). Jiangsu Agric Sci. 2000;21(4):37–40.
11. Zhang W, Wang Y, Wei B, He L, Zhang L, Liang J, Li M, Zhang L. Efficacy of chitooligosaccharide aqua in controlling wheat rhizoctonia root rot. Hubei Agric Sci. 2008;47(4):433–4.
12. Liu X, Du Y, Bai X. Relieving effects of oligoglucosamine on the inhibition induced by deoxynivalenol in wheat embryo cells. Acta Bot Sin. 2001;43(4):370–4.
13. Ben-Shalom N, Ardi R, Pinto R, Aki C, Fallik E. Controlling gray mould caused by Botrytis cinerea in cucumber plants by means of chitosan. Crop Prot. 2003;22:285–90.
14. Ma Q, Sun H, Du Y, Zhao X, Shang H. Induction of oligosaccharide to ultrastructure of cucumber resistance to powdery mildew fungus. Acta Phytopathol Sin. 2004;34(6):525–30.

15. Ma Q, Sun H, Du Y, Zhao X, Shang H. Effect of oligosaccharide on the resistance induction of cucumber against Sphaerotheca fuliginea. J Northwest Sci Tech Univ Agri For. 2005;33:79–81.
16. Xiao Z, Jiang X, Li X, Mo X. Preliminary study on the inducement of resistance against powdery mildew of pepper by oligochitosan. Hubei Agric Sci. 2009;48(3):617–9.
17. Xiao Z, Jiang X, Li X, Zhang S. Changes of activities of defensive enzymes in pepper leaves treated with chito-oligosaccharide and inoculated with powdery mildew. North Hortic. 2009;8:16–9.
18. Xu J, Zhao X, Bai X, Du Y. Effect of oligochitosan on controlling pepper phytophthora blight in field and on phytophthora capsici in vitro. Chinese Agric Sci Bull. 2006;22(7):421–4.
19. Yin H, Wang W, Lu H, Zhao X, Bai X, Du Y. The primary study of oligochitosan inducing resistance to sclerotinia sclerotiorum of Brassica napus. Acta Agric Boreal Occident Sin. 2008;17(5):81–5.
20. Xu Z, Li L, Li C, Qi J. Study on zhongshengmycin and amino-oligosaccharins to Fusarium oxysporum on watermelon. China Vegetables. 2003;3:10–2.
21. Pospieszny H, Atabekov JG. Effect of chitosan on the hypersensitive reaction of bean to alfalfa mosaic virus (ALMV). Plant Sci. 1989;62:29–31.
22. Pospieszny H, Chirkov S, Atabekov J. Induction of antiviral resistance in plants by chitosan. Plant Sci. 1991;79:63–8.
23. Shang W, Zhao X, Du Y, Shang H. First report on oligosaccharide induced resistance to plant virus. J Northwest Sci Tech Univ Agri For. 2005;33(5):73–5.
24. Shang W, Wu Y, Zhao X, Du Y, Shang H. Induced resistance to TMV multiplication in tobacco with Chito-oligosaccharides. J Northwest Sci Tech Univ Agric For. 2006;34(5):88–92.
25. Shang W, Wu Y, Zhao X, Du Y, Shang H. Effect of chito-oligosaccharide Induction on long-distance movement of TMV in tobacco. Acta Bot Boreal Occident Sin. 2006;26(9):1759–63.
26. Shang W, Wu Y, Zhao X, Du Y, Shang H. Inhibitory effect to TMV-CP gene expression in tobacco induced by chito-oligosaccharides. Acta Phytopathol Sin. 2007;37(6):637–41.
27. Shang W, Wu Y, Shang H, Zhao X, Du Y. The inactivating effect of chito-oligosaccharides on TMV particles in vitro. Chinese J Virol. 2008;24(1):76–8.
28. Shang W, Wu Y, Zhao X, Du Y, Shang H. Changes of defensive enzymes and PR-1a gene expression of tobacco induced by chito-oligosaccharides. Acta Phytopathol Sin. 2010;1:99–102.
29. Zhao X, Du Y, Bai X. The field experiment of oligochitosan controlling plant virus disease. Chinese Agric Sci Bull. 2004;20(4):245–7.
30. Rabea EI, Badawy ME, Stevens CV, Smaggbe C, Steurbaut W. Chitosan as antimicrobial agent: applications and mode of action. Biomacromolecules. 2003;4(6):1457–65.
31. Kuang Y, Peng H, Ye G, Qin C. Effects of oligochitosan on cold resistance of eggplant seedlings. Northern Horticulture. 2009;9:14–17.
32. Zhang W, Sui X, Xia W, Zhang Y, Lin C. Preparation of chitoligmer and its application to cucumber growth. Journal of Functional Polymers. 2002;15(2):199–202.
33. Guo W, Zhao X, Du Y. Effects of oligochitosan on the growth and photosynthesis and physiological index related to photosynthesis of tobacco seedlings. Plant Physiol Commun. 2008;44(6):1155–7.
34. Guo W, Zhao X, Du Y. Effects of oligochitosan on the seed germination seedling growth and photosynthetic characters of cucumber. Chinese Agricult Sci Bull. 2009;25(3):164–9.
35. Li Y, Wei L, Wang Q, Li H, Zhao X, Hou H, Du Y. Effects of oligochitosan on photosynthetic parameter of Brassica napus L. leaves. Chinese Agricult Sci Bull. 2010;26(2):132–6.

Printed in the United States
By Bookmasters